Black Fire

Black Fire

Portrait of a Black Memphis Firefighter

Robert J. Crawford Sr.
with Delores A. Crawford

Charleston   London

History
PRESS

Published by The History Press
Charleston, SC 29403
www.historypress.net

Cover image: Deputy Chief Robert Crawford. *Photograph courtesy of the Memphis Fire Department.*

Originally published 2003
The History Press edition 2007

Manufactured in the United Kingdom

ISBN 978.1.59629.328.1

Library of Congress Cataloging-in-Publication Data

Crawford, Robert J., 1931-
  Black fire : portrait of a black Memphis firefighter / Robert J. Crawford
Sr. with Delores A. Crawford.
    p. cm.
  ISBN 978-1-59629-328-1 (alk. paper)
  1. Crawford, Robert J., 1931- 2. Firefighters--Tennessee--Memphis--Biography. 3.
African American firefighters--Tennessee--Memphis--Biography. 4. Memphis (Tenn.)-
-Biography. I.
Crawford, Delores A. II. Title.
  TH9118.C73A3 2007
  363.3708996'073--dc22
  [B]
                        2007021483

To my family and to the three Pioneers of May 20, 1874,
and to the twelve Pioneer Black Firefighters of 1955.

Contents

Acknowledgements

I am deeply grateful to those who have provided assistance and encouragement.

My special thanks to Norvell E. Wallace, who was there from beginning to end.

Thanks to Karla Foster, André Foster, Marian Pride, Dr. Nellie Tate, William Adelman, James Mack Willis, Gerald Taylor, Annie P. Able, Mazella Flowers, Mable Barringer, Virgie Watson and Lawrence Lum, Stephanie Turner, Daniel Gregory, Loretta McBride, and James O. Rodgers.

To Donald Strickland, Memphis Room, Memphis and Shelby County Public Library. Thanks for your enduring assistance with the research.

Without the love, patience and faith of my wife, Delores, this book would never have been started and finished.

Tribute To A Negro Fireman

I heard the engines' clanging gongs,
    A block or two away.
And then I saw the raging fire,
    Dark smoke and waters' spray.

I saw the shiny ladder
    As it reached up to the wall.
And then I saw him climbing,
    Climbing upward, toward the call.

His black hands gripped the ladder,
    Which he climbed with sured pace.
The smoke engulfed his body,
    Flames danced about his face.

"I can't hold on! Please help me!"
    A youthful voice, a pleading cry.
"Hold on! Hold on! I'm coming!"
    Was his firm assured reply.

The roof began to crumble.
    The building's end was near.
Those below began to scatter
    At the sound which filled their ears.

His dark face was gripped with horror,
    His mind was seized by fear.
As he reached the fiery window
    He heard—"Swing the ladder clear!"

In that next heroic moment
    As I closed my eyes to pray
A black hand grasped the child
    And lifted him away.

There atop the ladder
    Clearly seen by every eye,
Were the fireman and the child
    Dark silhouettes against the sky.

He was grimy, hot and haggard
  As he stepped down to the ground.
A cheer arose—he smiled,
  But he never turned around.

When a reporter asked his name,
  I heard him quietly say—"NO NAME, PLEASE!"
"Compared to bigotry and other barriers I've overcome,
  This was an easy day."

—James O. Rogers
Author of *Blues and Ballads of A Black Yankee*
Exposition Press, 1965

Chapter 1

Brief History of Memphis's First Pioneer Black Firefighters

You may not know about the city of Memphis's very first black firefighters unless you read Joe Walk's book, *History of African-Americans in Memphis Government*. You may even have read the account of the first black firefighters in the April 14, 1874 edition of Memphis's *Daily Avalanche*.

Memphis held its general council meeting on April 14, 1874. Joseph Clouston, a black councilman, moved that the chief of the fire department put ten colored men on the force. The council overwhelmingly approved this proposition. However, that resolution was never carried out. Mayor John Loague promised during his campaign "to know no nationality or color in appointing to office." On May 20, 1874, the mayor appointed three black men, Joseph Luster, Peter Mitchell and Andrew T. Trigg to the Memphis Fire Department. The mayor later advised Chief Michael McFadden, "Three…are colored." Chief McFadden refused to recognize the new firemen. The issue of the appointment bogged down with maneuvering. Chief McFadden stated, "Only the Fire Committee of the General Council, with the chief's concurrence, could hire firemen." They did not pay Luster, Mitchell and Trigg for time worked for the fire department.

The three rejected firemen reportedly functioned as horsemen for a hand-drawn and hand-operated pump. They performed routine duties in their quarters and on the equipment. After only eight days with the fire department, Luster, Mitchell and Trigg were jobless. They never fought a fire.

Eighty years later, in March of 1954, Commissioner Claude Armour announced plans to hire black firefighters. Plans announced in the *Commercial Appeal* were not without pressure from the black community and the NAACP. Like the three black firefighters before them, the new firefighters had their work cut out in dealing with rejection and unfair legalities of the administration.

Fire Museum of Memphis. Restored antique fire equipment. Learn what to do in a fire. Children and parents are educated on fire safety awareness. *Photograph courtesy of Robert Crawford.*

Their experiences promised to be different than the "original three of 1874." The new pioneer black firefighters of 1955 persevered. We fought for full recognition and acceptance by the City of Memphis Fire Department. This fight for full recognition and acceptance was the most unforgettable experience of my life. It was, in reality, an experience in life itself.

Chapter 2
Growing up in Memphis

I was born in Memphis, Tennessee in 1931, the year that Ida B. Wells-Barnette died. I am an only child. As far back as I can remember, the neighborhoods I lived in all had one thing in common; every house we lived in was the last house on a dead end street. I was nearly five years old when we lived in a shotgun house at the dead end of Roper Alley. I remember the large, black, potbellied stove in the front room. I learned that they tore down our house to build Lamar Terrace, a housing project for white people only. I remember the grocery store on the corner. Mother would send me to that store with a note and a dime to purchase a ten-cent soup bone or a can of Steamboat Syrup. The ice cream parlor was across the street from the corner grocery store. I stood looking at the ice cream parlor, wishing I had a nickel for an ice cream cone.

I was still pre-school when we moved from Roper Alley to the dead end of Bismark Street in South Memphis. Our green and white house was a six-room clapboard with a long porch across the front. It had large front and back yards. We had a vegetable garden that included greens, okra, corn, tomatoes, watermelons and cabbage. Apple, pear, peach and fig trees grew naturally in our back and side yards. I attended Florida Street School from first to third grade. Mrs. Radcliff, my favorite teacher, had eyes shaped like Bambi's. Her smile was always warm and friendly. When she read fairy tales I held onto every word. I swear at times she made the characters come alive and leap off the pages and into my imagination. My father was the first and most influential teacher in my life. He was small-framed, five feet ten inches tall. His sister, Lottie Thomas, of Holly Springs, Mississippi, raised him. Aunt Lottie was a tall woman with a Southern cultured soft voice. She taught at the old one-room black school in Holly Springs. She raised two

The early years (1932).

sons and four daughters. My father, Lee Gusta Crawford, became a Baptist preacher. Naturally, I have many cousins on my father's side. Leaving his large family in Mississippi because he had no desire to farm, my father came to Memphis to look for work. He found a job as a porter at Methodist Hospital. He also founded Old Salem Baptist Church. The church was located at 694 Scott Street in Binghampton, a small community in northeast Memphis. My father met and married Lee Anna Macklin of Cordova, Tennessee. I have lived in Memphis my entire life.

Among my earliest memories is my father saving "Uncle Dad" from an injured hog. I never knew why they called my great-grandfather "Uncle Dad." Uncle Dad and my great-grandmother, Neely, lived with us. It was a cold, cloudy Memphis winter morning. An earlier short, sharp shower of rain made the air crisp and chilly. This kind of day was considered a good day for killing hogs. Uncle Dad sharpened the butcher knife on a whetstone with long deliberate strokes. Momentarily, he'd stop look up at my dad and say, "I believe these hogs know dey gettin' ready to die." My dad, a quiet man, mumbled a response. Leaning against the fence post, he watched Uncle

Dad test the sharpness of the knife's blade. It was not sharp enough, and Uncle Dad placed the knife again on the whetstone and continued to rub the blade against the stone. I stood close to my father. I listened to metal against stone *schwack, schwack, schwack* as he glided the knife over and back, over and back, across the sharpening stone. It must be sharp enough to cut cleanly through the hog's throat. He had to make a deep, wide cut and make it very swiftly. He couldn't dally around with a wounded hog. Uncle Dad grabbed the hog around the neck. He plunged the long razor-sharp knife's blade deep into the hog's throat. Blood streamed from the wounded animal like water jets from a geyser. The hog opened its mouth. He screamed in deadly guttural agony. The raging wild hog dragged Uncle Dad around the yard. My dad picked up a nearby sledgehammer. Uncle Dad and the hog came back around to our side of the hog pen. My father lifted the sledgehammer, aimed and came down with a swift blow to the hog's head. The hog lay lifeless. When it was all over, Uncle Dad, bloodier than the dead hog, came out with a few bruises and cuts. My dad said Uncle Dad held on to the hog because Uncle Dad needed to finish the job. He said that Uncle Dad knew what he was doing by holding onto the animal. A wounded hog, my dad said, is a dangerous hog. Uncle Dad was stiff, sore and achy for a good while after that encounter. He was lucky that my dad had a good aim.

My mother, Lee Anna Crawford, an only child, was born in Cordova, Tennessee, a small town in the country, adjacent to Memphis proper. Her mother died when she was very young and her grandmother raised her. They were sharecroppers. They lived in the country until my mother finished school. Besides working in the field, she did domestic work. My mother, great-grandmother and Uncle Dad moved to Memphis where my mother continued domestic work. When I was thirteen years old, she worked at Kennedy General Hospital in the dietary department.

My mother was small, five-feet-two and chubby, with an almost perfectly shaped round face and brown eyes. She had thick black hair that she wore braided across the top of her head like Rosa Parks. She had a beautiful smile that she used only occasionally. My mother was a very serious woman. When I think about her, I see her sitting near my great-grandmother, and the two of them are mending socks and work pants. I also remember the prayer meetings held at our house. You could hear the church folks a block away singing the old spirituals like "I love de Lawd, He heard my Cry." The song had a simple refrain, so that everyone could join in and sing it. After the singing and praying were over, my mother served a delicious meal that she and Aunt Lottie spent the entire day preparing.

Parents had their own ways of doling out discipline. While my father was the philosophical disciplinarian, my mother was the physical disciplinarian.

I remember playing hooky from Sunday school. Mrs. Flossie, a neighbor, told mother she saw me shooting marbles with other boys that morning. My mother made me get a peach tree switch. The last few licks ended with, "and you better not steal nothing, either." She'd end all whippings with these words. I wondered at the time what playing hooky from Sunday school had to do with stealing.

My parents and great-grandparents were nurturing and loving. They encouraged me to make decisions at an early age. They were hard-working people who lived through the Depression. Naturally, they raised me to work hard. One thing that my parents required was that I save a part of all my earnings. My great-grandmother kept my savings in a bag tied around her waist. She did not believe in banks.

When I was eight years old, my family moved from the dead end of Bismark Street in South Memphis to the dead end of Philadelphia Street in Orange Mound. The street dead-ended at the railroad tracks. If a train was across the tracks, we crawled under the train or walked around it. The house at 1285 Philadelphia Street was the largest house we ever had. More important, it was the first house that my parents would own. The house had imitation brown-brick siding, with a living room, three bedrooms and a large kitchen. The grounds were more than enough for my great-grandmother to raise chickens and plant a garden. I walked from our house to Melrose School.

I was nine years old when I started to work at Harold's Sundry as a delivery boy. One night, as I was returning to the store after making a delivery, an angry-faced man accosted me. He grabbed the handle bar of my bicycle and demanded, "Gimme yo' money." Although the three dollars that I'd collected was a lot of money in 1940, without arguing, I honored his request politely.

When I returned to the store, I told Mr. Harold what happened. He said nothing. From that I learned Mr. Harold trusted me. I was the only delivery boy that he allowed behind the counter and in the back room where he kept beer and sodas. He also kept buckets of change there. He'd ask me to stay after closing to restock the drink boxes. He paid me with a five-cent Spur Cola, which was a big thing for me. My salary was fifty cents per night. I had no need nor did I think about stealing from Mr. Harold. I was quite happy with the Spur Cola and I looked forward to my salary. But most of all, my father's sermons about the evils of temptation and my fear of my mother's whippings were more than enough to keep me on the straight and narrow. I dared not even think of touching Mr. Harold's money. The man who robbed me under the Boston Street Viaduct at Southern Avenue that summer night in 1940 apparently didn't have a father like mine and I doubt seriously that he had a mother like mine.

Memphis Central Station. Completed in 1914, the astonishing architectural structure was initially the Illinois Central Railroad Company. It was used for passengers and offices. Presently the stately building is used by Amtrak as a passenger terminal. *Photograph courtesy of Robert Crawford.*

Stax-Soulsville USA is the only soul museum in the world that spotlights major soul musicians who recorded there with exhibits, films, artifacts, memorabilia and galleries. *Photograph courtesy of Robert Crawford.*

My orientation to cultures other than my own came in my formative years. I am thankful they were positive ones and I thank my parents for the part they played in this. These earlier contacts prepared me for negative contacts that I would face later in my life. My mother was good at finding employment for me. She worked near Garbarini's WEONA, a grocery chain. She shopped at the grocery store for the families for whom she worked. Since her bus stop was near the store, my mother decided to talk to Mr. Joseph Garbarini Sr. about giving me a job there. I started to work at the store when I was ten years old. Mr. Garbarini was about five feet, six inches tall with black hair and a grandfatherly smile. Mr. Garbarini's son, Joseph Jr., was three years older than I. We argued about things I can no longer remember, but we never physically fought. The Garbarinis were Catholic and Joe attended Christian Brother's High School in Memphis. I liked Mr. Garbarini because he was a kind man. Other people seemed to like him too, because everybody that came into the store would laugh and talk to him. I never heard him say an unkind word to anyone, about anyone or against anyone.

I rode my bike from Orange Mound to WEONA's store. My friend Charlie "Charlieboy" Taylor and I worked at the store together until he quit. We were the same age. I was short and skinny-framed while Charlieboy was big-boned and taller than I. We attended Melrose School. We were not always friends. When I first started at Melrose School, Charlieboy and I fought. He said, "You pushed my sister down."

"I didn't push your sister," I said. Charlieboy pushed me and the fight was on. I had Charlieboy down and was wearing him out. Charlieboy had lots of brothers, sisters and cousins. His family, the Bridgeforths, lived all over Orange Mound. Each time I was on top beating Charlieboy, one of his brothers or cousins would roll us over so that Charlieboy was on top beating me. While Charlieboy was on top beating me, I looked up and saw Lawrence, an older boy who lived down the street from me. I was convinced he would help me. As it turned out, Lawrence was also Charlieboy's cousin. I got the worst of the fight that day.

Charlieboy worked on weekends and I worked weekends and evenings after school. One of our duties was to pick up milk bottles from customers and return them to the store. I dutifully picked up my bottles. On the way back to the store, I met Charlieboy. He had a threatening expression. He didn't want to pick up milk bottles; he wanted to take the bottles I'd gathered. I refused and when he tried to take my bottles we fought. His brothers, uncles and cousins were not there to protect him. I beat the snot out of him. I kept thinking about the licks he'd given me. I showed no mercy. I beat him so bad I was a little ashamed, but I didn't feel sorry for him. After that fight I had no further problems with Charlieboy. He and I remained friends after he left the store.

You could find almost everything you needed in the neighborhood where I worked. The grocery store, the hardware store, fish market, the dry goods store, dry cleaners and everything were there. The best thing about all of this was that people knew each other. I got to know customers, too. During the week, customers ordered groceries or came by the grocery store during the day and I delivered their groceries after school. The local trade was especially busy during the holiday seasons. Delivering groceries was exciting, especially in November and December when customers tipped more and gave me Christmas presents.

I delivered groceries all over the area including the nun's convent. The convent was part of St. Augustine Roman Catholic Church and School. It was one of two black Catholic facilities in Memphis. The other black

Lemoyne-Owen College. The historical black liberal arts teaching institution's history dates back to 1882 when a nurse from Chicago, Lucinda Humphrey, began teaching freed slaves the fundamentals of reading. *Photograph courtesy of Robert Crawford.*

Catholic Church and School was St. Anthony at 327 Concord Street in North Memphis. St. Augustine was at 903 Walker Avenue near Lemoyne-Owen College. The Baptist Church was all I ever knew. Since I was young and impressionable, my curiosity grew regarding the nuns, the Catholic school and the church.

When I was not delivering groceries I stocked shelves. During this time I vicariously took part in grownup conversations about World War II. "When You Wish Upon a Star," played on the store radio between news of the war. Women cried openly when the Japanese attacked Pearl Harbor. Most of the men talked about President Roosevelt and the war declared on Japan. There were controversial thoughts about what was being done to the Japanese American population. "Remember Pearl Harbor," was the cry of the man on the street. President Roosevelt, with his fireside chats, was on the radio often talking about the war. We had no electricity. My parents listened from our battery-operated radio. A long wire from the radio led to the window and dangled outside. We had good reception.

One day Mr. Garbarini and I were alone in the store. He stopped his work and approached me as I was rotating the canned foods to the front of the shelf. I looked up and acknowledged his presence.

"Robert," he began, "Would you like to attend St. Augustine School?" "Yes sir," I answered. I am sure I expressed excitement in my answer and on my face. "I'll speak to your parents," he said.

I contemplated what my parents would say about my going to a Catholic school. I imagined various scenarios in my head about what the possible answer would be. I hoped they would say yes.

A few days later my parents talked to me about changing schools. "Robert," my father said, "Mr. Garbarini said you want to go to Catholic school."

"Yes, sir," I said. I was sure he'd say no since he was a Baptist preacher.

"Your mother told Mr. Garbarini that we are not able to pay. Mr. Garbarini said he would take care of the tuition."

I was happy that my parents decided to let me go to Catholic school because I'd become curious about the school. The nuns were warm and kind when they visited the store and when I'd deliver groceries to the convent. They knew me by name and often spoke very kindly to me. They heightened my curiosity about the school. I started St. Augustine School in the middle of the school year. I was eleven years old. Math was my favorite subject. I still like working with numbers today. The nuns were stern teachers. Sister Mary James was one of my favorite teachers. She and all the nuns were strict and demanding, especially when it came to good penmanship.

I had little time for socializing, except when I attended Boy Scout meetings once a week or when I went to Boy Scout summer camp. My work schedule kept me busy after school and on weekends. I remember playing football during lunch. I broke my kneecap. The older boys took me to Dr. Roulhac's office. I ended up in the emergency room at John Gaston Hospital. My injured knee was painful, but I did not cry. They hospitalized me and I was out of school for two weeks. I returned to school on crutches and I enjoyed the attention from other students during my recuperation.

Six weeks later I returned to work at Mr. Garbarini's store, where I would work for the next seven years. I became close to the family and I knew members on both sides of the family. I'd spend part of Christmas Day at the Garbarini's home. They always made me feel like family. I remember going fishing with Mr. Garbarini. I never caught any fish. Neither did Mr. Garbarini, at least on the trips I made with him. I am sure that Joe Jr. remembers the time that Mr. and Mrs. Garbarini, Joe, his girlfriend, Jean, and I went on a fishing trip. Now there were two boats and I had to choose between the two boats. I chose to boat with Joe and Jean. I recall the taut look on Joe's face. I think my intrusion spoiled his fishing trip that day because he wanted to be alone with Jean.

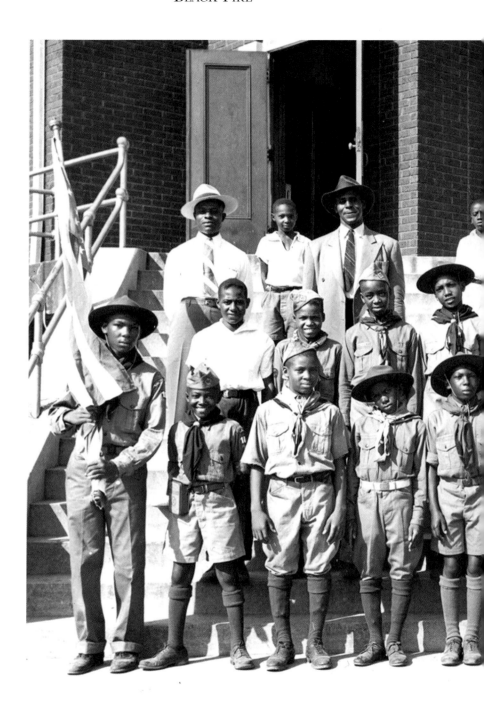

Boy Scout Troop #104, Orange Mound (1940s). Robert Crawford is the second person from the left in the second row; Charlieboy is third from left in the front row. *Photograph courtesy of Boy Scout Troop.*

Church was an all-day affair for our family. Traditionally, we were up before daybreak. On this winter Sunday morning, daylight came long before the sun. We could not tell if it would be a cloudy or a sunshiny day. Sunday mornings my grandmother always made biscuits that stayed light even after soaking them in the butter she'd churned. The country bacon left its aroma throughout the house. I felt it was a good time to bring up the subject. "I want to join the Catholic Church." I directed my statement to my parents and to my great-grandmother.

Daddy spoke first. "I would rather you join the Catholic Church than no church." My father's agreement led me to believe that he sensed my developing lack of interest in staying at church all day. My mother's strict rule against my playing in my Sunday clothes and the long round-trip bus ride from Orange Mound to Binghampton must have helped him make the decision. I really felt that he was afraid I would lose all interest in the church. I know he did not want that to happen. " As long as you go to church, it's all right with me." My mother agreed. My grandmother approved with a nod and a smile. I returned the smile. The kitchen finally lit up as sunrays streamed through the kitchen window.

I was an altar boy from the sixth grade until I finished high school. The thrill of my boyhood was serving at mass. I carried the crucifix on special occasions. I found the Catholic Church different from the church in which I grew up. The priest dressed in colorful vestments. Each colorful vestment represented a season in the church year. The church always smelled of incense. People entering blessed themselves with holy water and made the sign of the cross before entering and upon leaving the church. They genuflected, or kneeled, before the altar and knelt to pray after entering the pews. The church was usually crowded at Easter and Christmas midnight mass with people who attended church at these special times only. Austere Christmas trees stood majestically, while white candles flickered on the altar. Christmas decorations included bright red bows attached to each pew. The priest and altar boys dressed in white for the occasion. Serving as altar boy was absolutely the highlight of my boyhood.

By the time I was in high school, I had saved enough money to buy a motorbike. At school, I was senior class treasurer. Senior treasurer was not the only title I held in high school. I was labeled "Father's left-hand man." Our priest, Father Bertrand Kock, came to Memphis in 1937 and founded St. Augustine Church and School. He was stocky and walked with a military gait boys tried to imitate. "Father Bert" knew that he could depend on me to serve daily masses and to run church errands. For being a dependable, faithful server of mass during school and Christmas holidays, I received a pair of skates. We had no sidewalks in Orange Mound, so I lent my skates

St. Augustine Catholic Church senior class altar boys Richard Thompson, Edward Flagg, Bobby Westbrooks, Robert Crawford, Alexander Deloach, Edward Wilhite and Linwood Chambers. *Photograph courtesy of St. Augustine Catholic Church.*

to my classmate, Warren Welch. I'd see him skating when I rode the bus to school. Warren wore my skates out.

When I served as altar boy I felt especially close to God. The adornment of the church's altar, carrying the crucifix and lighting the altar candles gave me a saintly feeling. However, I was no saint by any stretch of the imagination. I had a busy work schedule and therefore had little time to get into trouble. Most of those "little times for trouble" occurred during school hours. My buddies in crime and I made a point of sitting in back of our classrooms. We shared mischief during school. We shared cigarettes in the bathroom. We also shared bad words, teased girls and occasionally fought

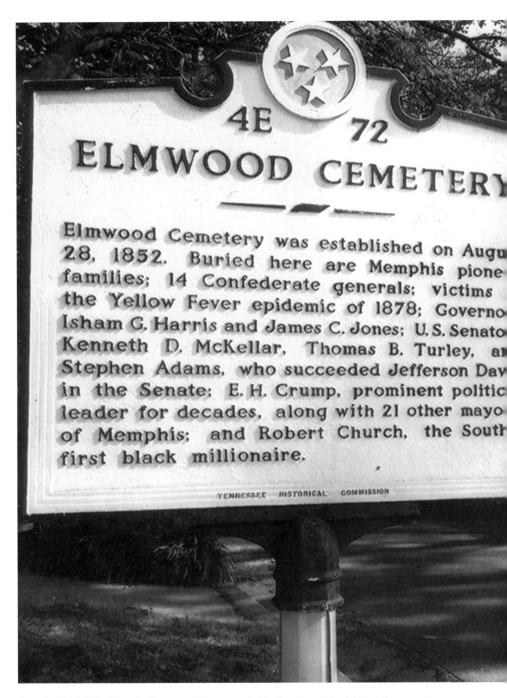

Founded in 1852, Historic Elmwood Cemetery is Shelby County's oldest active cemetery. It is on eighty landscaped acres with many Victorian statues and monuments. *Photograph courtesy of Robert Crawford.*

Left: Sister Mary Janelle, BVM—Robert's favorite teacher. *Photograph courtesy of Robert Crawford.*

Opposite: The Poor Clares of Memphis gathered from many parts of the United States by following a personal and unique call from God. In order to follow Jesus in poverty, chastity and obedience, the sisters have embraced a lifestyle that is contained within the walls of the Monastery of St. Clare. *Photograph courtesy of Robert Crawford.*

with other boys. On this day while Sister Janelle was out of the classroom, we shared rubber bands. Bobby Westbrook was a paperboy. Therefore he carried in his pocket a supply of rubber bands he used to wrap newspapers. He passed a rubber band to each of us. We folded notebook paper in small wads and placed the wads of paper in the slingshot fashioned over our thumb and second finger. We drew and aimed in unison. My wad of paper landed squarely at the back of Alexander's head. He grabbed the back of his head while uttering an expletive as he felt the sting. He jumped out of his seat and focused angrily at my buddies and me.

Sister Janelle looked around. She never called us by name but simply said, "Those who shot paper stay after school." My buddies and I shared after-school time listening to reasons we should not misbehave in or out of school.

My experience and knowledge grew as I performed tasks for the church. In the process of running errands for the church I met different people, visited new places and learned interesting things about my environment. For example, I remember going to the cloistered nuns' convent in Frayser, a small community north of Memphis. The nuns made the altar bread for all Catholic Churches at the Poor Clare Convent. The nun spoke through a

door that had a small carousel attached. She placed the bread on the door's carousel. I never saw the nun's face.

Sister Mary Janelle was my favorite high school teacher. Whenever Sister Janelle was angry, she'd point her crooked fourth finger at you and you knew what she meant or you'd find out real quick. She gave me a Catholic Pocket Missal that she inscribed with "Robert, may you always be as near to the altar." I still have the missal.

My world continued to revolve around school, church and work. Most of my socializing continued at school because of my work schedule. Thomas Ray Moore, my friend, attended St. Augustine School. Thomas Ray was tall, friendly, courteous and a perfect gentleman. We worked at Duke's Drive Inn on Summer Avenue during summers and weekends. Catholic school used to have dances. I never had time to learn the popular dances, but I agreed to go with him and Barbara to the dance at St. Anthony Catholic School in North Memphis. Barbara introduced me to Delores Anderson.

It was 1948 and the girls wore ballerina shoes and full skirts. I can see Delores now with skirt in full swing and moving to every beat of Count Basie's "One O'clock Jump." I kept trying to get close to this girl who was a good dancer. When they played "Nature Boy" by Nat King Cole, I asked

The shotgun shack at the W.C. Handy House Museum, once the Memphis home of W.C. Handy, "Father of the Blues," is where he wrote "Beale Street Blues" and the "Memphis Blues." Collections of memorabilia, artifacts, old photos and commentary by the museum guide are highly informative. *Photograph courtesy of Robert Crawford.*

her to dance, but by the time we walked to the dance floor, the song was over. She looked at me and we both laughed. They started playing a fast number before we left the dance floor. She looked at me sideways. I told her that I couldn't dance. She offered to teach me to dance. I knew then that she liked me too. The next time they played a slow number, I danced with Delores.

I made time to see Delores. I'd pick her up and she would accompany me on errands for the church. Every time I got the chance I'd visit her. Sometimes, Ray, Delores and I met at Barbara's house. She always had the latest records. We laughed a lot while Delores tried to teach me the latest dance. She said, "You just have no rhythm or you're not interested in learning to jitterbug." I never learned the latest dances, but we did get to know each other.

I was a graduating senior in 1950. It was a happy time of my life. I had two jobs and an old 1938 Plymouth that I constantly worked on to keep it running. I was the only student who owned a car at St. Augustine Catholic School that year. I worked from 4:00 p.m. to 6:00 p.m. Monday through Fridays. I also worked from 8:00 a.m. to 8:00 p.m. at the grocery store on Saturdays. My second job, at Duke's Drive Inn on Summer Avenue, was on Fridays from 7:00 p.m. to 1:00 a. m. and Sundays from 6:00 p.m. to midnight. In my life, since my first job, I cannot remember when I was without two jobs. I saved part of my earnings and I helped my parents with household expenses. For a teenager, I did well financially.

Delores continued the dancing lessons, but my feet refused to cooperate with either of us. She was my first love. We spent as much time together as possible. Delores had nine sisters and three brothers. It was a long time before I learned the names of all of her sisters and brothers. We were married in 1950. Only Barbara and Ray knew. We kept it a secret because I planned to volunteer for the army. The army would not accept married men on a volunteer basis. After I enlisted, we told our families about our marriage and showed them our Hernando, Mississippi, marriage license. The Catholic Church said the marriage was not valid because a Catholic priest did not perform the ceremony. Father Capistran Hass married us in the priest's house at 917 Walker Avenue before I was shipped overseas. Barbara and Thomas Ray married the same year. Thomas Ray was drafted into the army.

Chapter 3

Leaving Memphis

Although I was doing well working two jobs, it was not enough to pay for college. I realized that college work is demanding. Therefore, I volunteered—along with three of my classmates, Earnest Morris, Edward Wilhite and Richard Thompson—for service prior to our graduation from high school in June. The armed service was the next best place to go if you could not afford college. Morris and Thompson joined the Air Force; Wilhite and I joined the army. The recruiter assured me that I would attend diesel mechanic school following basic training. I felt great about my decision to enter the armed services.

Uncle Rex, the best shade-tree mechanic in Memphis, taught me all he knew about cars. However, I wanted to learn everything about diesel engines, so I planned to get diesel mechanic training in service. I was excited about my first trip outside Tennessee. I traveled in Tennessee and some parts of Mississippi because I have relatives in these places, but I never ventured north of the Mason-Dixon Line.

New friends that I met in Fort Knox, Kentucky, during basic training became old friends as time passed. My new friends and I discussed basic training and various reasons why we volunteered for service. Most of us joined for occupational reasons. We planned to use our new technical skills for a better life following our tour of duty. Jacob Johnson, whom we felt was the most ambitious, wanted to be a career soldier. All of us agreed the U.S. Army was our college. We didn't have to pay tuition, which is why we were there in the first place. I had my dreams all wrapped up!

I met Thomas Caleb in Fort Knox, Kentucky during basic training. Thomas, a well-read guy, was from Philadelphia. We hung around the base together on weekends. Caleb was a nice, easygoing guy with a great sense of humor. We remained in the same company in Korea.

In July 1950, we were halfway through basic training. Suddenly everything changed. Our officer shocked us one morning during training class. "All schools are closed. You are now training for combat in Korea," the officer roared. On June 25, 1950, the Korean War began when North Korea forces launched an invasion across the 38th Parallel into South Korea. We listened silently, our dreams shattered. We entered at peacetime and we end in war. I was assigned to an infantry company.

Seoul and Pusan are considered "special cities" in Korea. We arrived in Pusan in October. This was the end of summer in Korea. Like Memphis, Korea has four distinct seasons. The seasons have various moods. A spring thaw comes in mid-April and lasts little more than two months. Early spring northwesterly gusts bring swirls of golden dust from the Gobi Desert and a light rain. As summer approaches, humid southerlies vie for control. Spring drizzle becomes an occasional downpour by summer, June through October. July and August are the hottest months. Autumn, by far the most splendid time to be in the country, comes in October.

Telephone linemen, medics and a few other schools were open when we arrived in Korea at the Pusan Perimeter in October 1950. I wanted diesel mechanics so I took nothing and remained in infantry. They called my name, loaded us on trucks and carried us to a town in Korea. After marching Korean prisoners to Pusan, we set up a base company and supervised prisoners building their own stockade. This was a new company formed when we got to the rear or away from the combat zone. This was the first time that I saw black officers. There were two black captains and a colonel. Since it was a new company, I remained there, though I was not a military policeman at the time. Nevertheless, they assigned me to the outfit because after a certain length of time with a company you automatically became part of that company. This assignment later proved favorable in many ways.

My company was new and I had what the army needed. They were looking for men who had mechanical experience. Our company had trucks that were not operating. The best thing was that they were looking for anyone who knew anything about mechanics. I knew about mechanics and I was good at reading and understanding automobile manuals. I volunteered for mechanic in the motor pool. In the motor pool, I met Private Charles Quenzenberry, a white guy from North Carolina, and a black private, Marvin Hip, from Alabama. Quenzenberry was tall and brawny and he always held a cigar between his teeth. He and Hip, a thickset dark-brown-skinned guy, had been army buddies for some years. They had gone to mechanic school before the Korean War. They knew army vehicles. The

Robert with army
buddy Charles
Quenzenberry in
Pusan, Korea (1950).

two soldiers were impressed with my ability to observe, understand and apply skillfully what I learned. They mentored me, teaching all they knew about engines. Both were career servicemen with forty years of army vehicle mechanics between them. We were big buddies because motor vehicles were our common interest.

As a mechanic I made certain the trucks were in good running condition. I drove heavy equipment over some of the worst roads possible. The roads were narrow with rice paddies on either side. These paddies were soft and once a vehicle was stuck it was extremely hard to get out. The worst part was trying to get the vehicle out of the rice field, as you had to work in human feces. Driving in the city was not as difficult. I also functioned as dispatcher. I sent trucks out whenever the first sergeant needed them. In addition, I transported prisoners to various work sites and picked up supplies at the dock. I didn't know it then, but experience driving heavy equipment, as well as my mechanical skills, would be to my advantage in the future. In 1951 my company moved the prisoners of war to Koje-do Island, Korea's second-largest island. Thomas Caleb and I were still in the same company. However, Caleb developed tuberculosis while we were in Korea so he left early. We communicated until I lost contact with him. Later, Delores wrote that my friend Thomas Ray Moore had been killed in action in Korea.

My rank was E4 when I returned to the states in 1952. Six other black soldiers and I were assigned to Sandia Base in Albuquerque, New Mexico. Our company in Korea was integrated before we left. We integrated Sandia Base. Soon the six of us were working all over the base. The different areas I worked on the base included gate guard, regular MP duties, train patrol and town patrol. I was the first black soldier on train patrol and town patrol and the first black soldier on motorcycle patrol. Captain Kenneth Bennett was an officer who paid attention to his men. He noticed me working in various areas on the base. I saw him passing through where I directed traffic on the base. I impressed Captain Bennett with my performance and he nominated me for Soldier of the Month. This was an honorable award that meant a lot to soldiers on the base. I won the award for that month.

I remember being on town patrol and receiving information that a soldier had gone AWOL. He was being held at the recruiting station. The soldier asked to go to the bathroom. He escaped through the bathroom window. They called for assistance in tracking him. After receiving a description of his clothing and a general description of the soldier, my partner and I separated. I saw the soldier running. He was really moving fast. I yelled a warning to stop or I would shoot. He cooperated. I didn't cuff him at the wrist. I cuffed his right wrist to his left ankle and made him walk back to the patrol car in a limping position. I never actually drew my weapon because I had no intention of shooting him.

I finally attended a gas-engine mechanic school on Sandia Base in 1953. I liked Albuquerque and wanted to remain there because of the opportunity to join the police force. My military experience would have secured me a place in that department.

Delores and our infant daughter, Joan Christine, visited the base. We spent lots of time in the Sandia Mountains. She enjoyed the mountains, but she did not like living there. I felt that with a little persuasion she would have stayed. However, I vowed to do everything to make my wife happy. I came home because my family wanted me home. I had no idea then that one day I would regret leaving Albuquerque, New Mexico and its job opportunities to return to Memphis. Finding a job in Memphis in 1953 was extremely disappointing.

Chapter 4

When Opportunity Knocks

In March of 1954 the City of Memphis announced in the *Commercial Appeal* they wanted to hire "colored" firefighters. The headline offered almost every black male in Memphis the opportunity of a lifetime. There were wall-to-wall black males seeking the twelve positions. A year passed. The City hired no black firefighters. Black civic leaders and business leaders held meetings in the black community. The NAACP pressured the city leaders whose answers were "we can't find enough qualified blacks for the job." The application and selection process continued until July 1955.

Until the March 1954 announcement in the *Commercial Appeal*, the thought of becoming a firefighter never entered my mind. I was a year out of service. Already I contemplated leaving Memphis for Albuquerque, New Mexico. There I could work as a police officer because of my experience in service as a military policeman. At home the City had hired the number of colored policemen it wanted. There was no telling if or when they would accept applications again. I gave up the idea of becoming a diesel mechanic when I came home because auto-mechanic schools were segregated and none were open to blacks. I knew it was futile to pursue the idea in Memphis. I wanted the fireman's job because I had a family to support and I had bad experiences finding a decent-paying job in Memphis. For the first time in my life I dreamed of becoming a firefighter. This was the best offer for a job in Memphis because it was secure. The retirement benefits were good and the job offered hospital insurance.

When the ad for black fighters was announced, I was working three jobs. I worked at McDonald Brothers Appliance Warehouse and M.E. Carter Fruit Company. I also attended Moler Barber College under the G.I. Bill. I cut hair at a barbershop on Saturdays. However, my focus shifted to the Memphis Fire Department and becoming a firefighter.

The Memphis Zoo displays over four hundred species of exotic and endangered animals. Exhibits include Dragon's Lair, Cat Country, Primate Canyon, Animals of the Night and many more. *Photograph courtesy of Robert Crawford.*

Slave Block on Auction Ave at North Main Street. *Photograph courtesy of Robert Crawford.*

Fire Station #12 was across the street from where I worked at McDonald Brothers Warehouse. I talked to firefighters at that station. I was curious about the fire department. I asked many questions. They willingly answered my questions and often gave additional information about firefighting. The white firefighters were friendly, talked openly and gave me much information about the fire department each of the many times I visited the station.

In 1955 postage stamps were three cents, a cup of coffee cost ten cents, soda pops also cost ten cents and hamburgers thirty-five cents. We used colored restrooms and drank water from fountains set aside for blacks. We were assigned separate restaurants. We were granted Thursdays at the zoo and assigned riding in the rear of the city bus. This same year the practice of "separate but equal" facilities in public school systems was declared unconstitutional. In Memphis the NAACP, along with black city leaders, continued applying pressure on the city's leaders to make good on its promise to hire black firefighters.

Fire Commissioner Claude Armour made clear before they hired the first black firefighters that following a forty-day training period, black firemen would be assigned to Fire Station #8 at Mississippi and Crump Boulevard. It was also clear that black firemen would have white officers and that white firemen presently assigned to Station #8 would be assigned to other stations. The black firefighters would serve specific areas. Specific areas included Vance St. on the north, South Parkway on the south, Bellevue on the east and Third Street on the west. Blacks populated these areas. Most important was Commissioner Armour's promise that "the Negro firefighters would be under supervision of a white captain and lieutenant until experiences and eligibility requirements are met for promotion. Firemen must serve five years before being eligible to be lieutenant and another three years before he can be made captain." Getting those promised promotions would prove to be Herculean tasks for black firefighters.

With all plans finally in place, including housing black firefighters, transferring white firefighters, specifying areas black firefighters would serve and making various promises, the department became bogged down with other issues. Meanwhile black firefighter candidates crowded the small red City personnel office building on Adams Street across from the courthouse. Fire administration bogged down because, according to administration, only 42 of the 175 perspective firefighters completed and returned their applications. Commissioner Claude Armour said, "We had thought we would have enough to start a class. From past experience, we don't think we could get a good selection from the number who have returned applications." It seemed inconceivable that the administration couldn't find 12 qualified men out of 175 applicants.

The NAACP continued pressuring Commissioner Claude Armour and the city fathers to hire black firefighters. They did not release the City from its commitment. The organization pushed relentlessly until the department kept its promise. On March 21, 1955, Commissioner Claude Armour again announced plans to hire black firemen. In July 1955, a program was set up and the first black fire company was underway. A group of black civic leaders and business leaders conferred with Commissioner Armour on the plan to hire black firemen. Those black leaders included Professor Blair T. Hunt, principal of Booker T. Washington High School. He expressed appreciation "on behalf of our colored citizens for this forward step." Others present during the announcement to hire black firemen included Reverend S.A. Owens, pastor of Metropolitan Baptist Church; James Coger, merchant and president of the White Station Civic Club; James T. Walker, president of Bluff City Council of Civic Clubs; R.S. Lewis Jr., president of R.S. Lewis & Sons Funeral Home; Robert Wright, president and founder of the Orange Mound Civic Club; and Sam Qualls, president of Qualls Funeral Home.

Prior to the hiring of blacks, no test for qualification was required for a firefighter's job. The only thing required was the desire. However, black

Shelby County Courthouse. *Photograph courtesy of Robert Crawford.*

applicants had to pass a battery of preliminary tests, including agility tests, civil service examinations and physical and oral interviews with the City personnel director before we were appointed to the position of fire private. In the preliminary application, each applicant had to provide references. Then, whom one knew or to whom one was connected determined who got the interview. Mr. Robert Wright was a smart man. I knew in those days a good reference from Mr. Wright was essential.

I met Mr. Wright, a tall medium-brown-skinned man, after my discharge from the army. I was an angry young man at the time because I encountered rejection after rejection while looking for work. It was not because of my qualifications, because I was never asked about them. My color was the deciding factor. I remember the time I went to the Ford plant on Riverside Drive looking for work. Entering the building I asked an employee for directions to the personnel office. He was neatly dressed in beige slacks; a beige shirt and a brown tie. "I would like an application for a job," I said.

The neatly dressed stony-faced man politely pointed to an old black man behind a broom and said, "That colored boy over there has worked here for twenty-five years and we don't plan to hire another one."

When Opportunity Knocks

I went to the Army Depot, a federal institution, for an application for gate guard. The announcement for this job was posted on the board. I informed the personnel worker that I was recently discharged from the army as a military policeman. She stared at me and then stated, "We don't hire colored gate guards."

These and other related job-hunting incidents reinforced my anger for a long time. Mr. Wright was unaware that he helped change my attitude and gave me a new focus. Neither Mr. Wright nor I knew it then, but for me, my association with Orange Mound Civic Club was really a workshop for developing leadership. It pointed me in another direction at a time when I truly needed direction toward other areas of my life. I had learned by observing and applying organizational skills in the army. In addition, I learned organizational skills as I participated in Orange Mound community projects. Mr. Robert Wright and the Orange Mound Civic Club provided invaluable lessons in how and when to form relationships with leaders in and outside of the community. For example, working closely with Mr. Wright, I observed the way he contacted local leaders, NAACP organizations and leaders in other cities. I was unaware at the time, but I was learning to network. These observations proved to be valuable tools later in my life.

Mr. Wright, president of the organization, was a strong black, respected community and political leader. I contacted and obtained a letter of reference from Mr. Robert Wright. As we talked he explained how pleased he was that I was applying for the fireman's job. He said, "I have watched you work with the Orange Mound Civic Club and I like your enthusiasm, the way you observe a situation and then go on to complete the task. You are task-oriented and you will do well," he said.

At the end of our visit I thanked Mr. Wright for all that he taught me and for the recommendation. "In the end," he said to me," it's not who you know but it's the kind of person you are and the kind of performance you give that make all the difference in the world." As I moved through life, I had many opportunities to reflect on Mr. Wright's words.

The Garbarini family, who aided in my Catholic education, was especially supportive and wrote a letter of high recommendation. The priest and nuns also wrote compelling letters for me. They were pleased that I applied for the job.

Later, it was announced that only ten firemen were actually needed, with ten alternates. It seemed like an eternity between the time the announcement was made in the *Commercial Appeal* and the actual training for black firefighters began. I was aware that physical examinations, checking of police records and the actual training period itself would eliminate some of us. Commissioner Armour and Fire Chief John Klinck announced that twelve applicants were accepted for training out of several hundred who applied for the positions. Training was scheduled to begin the following Monday, July 11, 1955.

Chapter 5

Training

July 11, 1955, twelve black men began fire training. I was one of the twelve pioneer black firemen. The other eleven were William C. Carter, John Cooper, Richard Burns Jr., George Dumas, Leroy Johnson Jr., Floyd E. Newsum, Elza Parson, Murray Pegues, Carl Westley Stotts, Norvell Wallace and Lawrence E. Yates. We were the new black pioneers. We began training with odds against us.

C.L. Scott was chief of training of the Memphis Fire Department. Our immediate training instructors were Captain William Eubanks and Lieutenant E.B. Selph. Captain Eubanks was a clean-shaven, energetic and experienced firefighting officer. Lieutenant Selph was precise and always prepared. He was his own man. I would learn more about him in my firefighting experience. Both officers had military attitudes. The training was held at the Memphis Fairgrounds, where Memphis held its outdoor fairs, circuses and exhibitions. It was the fairgrounds where "coloreds" were allowed only on Thursdays and special days. However, since we were in training we were permitted on the grounds for that purpose. We trained from 8:00 a.m. to 4:30 p.m. We trained in our own clothes that were much like those any laborer wore. Most of us wore brown or green khakis while others wore blue jeans. Our shirts included anything from sweats to tee shirts. The instructors wore full dress summer khaki uniforms.

We had two sets of instructors, one for white recruits and one for black recruits. White recruits started training July 18, 1955. Although we had the same training, we trained in different areas of the fairgrounds. Following training, each evening the white recruits went to different fire stations for on-the-job training for firefighters. They were placed with seasoned firefighters. The colored recruits went to the "black fire station" at Mississippi and

After fire training. Mentor, R.S. Lewis *(standing)*, firefighters Norvell Wallace, John Cooper, Richard Burns, Leroy Johnson, Robert J. Crawford, Floyd E. Newsum, Elza Parsons and George Dumas. *(Sitting)* Lawrence Yates, Carl Stotts, Murray Pegues and William C. Carter. *Photograph courtesy of Earnest Withers Photographer.*

Crump Boulevard. The white lieutenant and white privates stationed there were moved to other stations for the night because of segregation. However, one white officer, Captain M.D. Baxter, and the white driver, G.B. Kuhn, remained at Fire Station #8. We remained at this station, segregated and isolated, for the next ten years. Training started at 8:00 a.m. We had physical training each morning. The classroom training included firefighting, first aid, nozzle pressures and water discharge of each nozzle used by the fire department. Some nozzles were hand-held and others were attached to fixed appliances. We learned the location and addresses of all firehouses in the city.

The outside training included ladder training, hose evolution and instructions on laying out hose. We raised 25-, 35- and 50-foot ladders. We advanced hose up the aerial ladders. We learned to tie knots used by firefighters. We learned to rappel buildings. All outside training was repetitious. We role-played acting as captain, lieutenant, driver, assistant driver and private on a regular firefighting company. We had written tests each week on that week's training. Becoming a Memphis firefighter was serious business.

The thorough training was not nearly as difficult as trying to live with crazy rules that say men can't talk or socialize. Since I was recently out of service, the recruits training separately seemed like a stupid idea and an enormous waste of time, energy and financial resources. Conditions in service had become much better before my discharge. Consequently, I looked for changes in Memphis. In basic training we had white company officers. Our drill sergeants were black noncommissioned officers. By the time I was discharged, black and white officers trained black and white recruits together. One particular incident sticks in my mind while we trained at the fairground.

The day was a Memphis kind—hot, humid and muggy. Flies light on your sweaty skin and stick to you. We had down time, or break, following lunch. We played softball with white recruits versus black recruits. We were playing ball together and enjoying our game. A car pulled to the curb. A stocky man exited and walked toward us. Heat waves around gave him a slow-motion appearance. He was the park police officer. Stopping the game and acknowledging his presence, we waited for him to approach us. There was silence. As he sauntered closer, he spat out a wad of brown tobacco. Wide perspiration rings circled under each arm. He was sunburned with eyes the color of a tarnished penny. In a drawling guttural accent he said, "colored and whites can't play together in this park." He broke up the game and stood there until the colored and white recruits went their separate ways. Satisfied that he'd gotten his message across, the policeman turned, trudged slowly to his car and left as he'd arrived in the flickering heat of a mid-day Memphis summer.

I learned early that firefighting is competitive. I felt that our training officers inadvertently positioned us in a competitive mode. The early August day was sultry, sweltering hot and humid. On this day our drills included rappelling the wall. With a rope tied to the top of the building, we rappelled down the side of the two-story building much like a mountain climber coming down a mountain. My body started to cool from the sweat of my clothes. I tasted the salt from my body. I saw water flowing from other black bodies and I felt tension as we competed *sub rosa* against each other. Nevertheless, I felt what made me more aware of the competition was not the idea of competing with each other. Rather, it was the trainer's attitude. I felt that he was surprised at our eagerness to excel. I knew we had to be exceptional in everything, including physical training and written examinations. We had to be the best. As time passed we were just that. We were fully aware that we were not welcome on the fire department. Talking it out with each other proved beneficial at times.

The two classes of recruits did not socialize freely. Therefore, black recruits were automatically drawn closer to each other. From our initial contact we began establishing a bond, because social mores dictated the separation of the races. We helped each other and we provided the support each needed to become a good firefighter. We were aware that one mistake probably meant an early end to a black firefighter's careers. Having no seasoned firefighters to relate to or to question, I relived the entire day's drills, going over little details in my mind. Black firefighters exchanged ideas on how to help each other master the drills.

After discussing events of our day's training, some recruits called it a night. I had developed a third ear. I learned to listen for the alarm as I slept. We were now making regular fire calls. I wanted to make fire runs because this was a new experience. On the other hand, I didn't want to make fire calls because I knew I'd suffer the next day from lack of sleep and the hot sun would drain what energy I had left.

I lay on the edge of the bed trying hard to concentrate on sleep. I looked at the five black privates asleep in the bed hall. The cots were lined against the wall like army bunks. I compared the situation at the fire station with my experience in New Mexico. Leaving Korea for stateside, I was among the first group of black soldiers to integrate the military police company at Sandia Base in Albuquerque, New Mexico. It happened the first night we spent on the base. We were assigned to a squad room where ten to twelve men slept. We went to sleep in a room full of soldiers. When we woke up the next morning, all the white soldiers had moved out and left the three black soldiers in the squad room. At most engines houses, the lieutenant shared the bed hall with the privates. At Fire Station #8 the driver did not sleep in the bed hall where black firefighters slept; he shared the captain's quarters. Unlike the situation on Sandia Base, the white firefighter apparently made his move before we arrived at the engine house.

Training continued rigorously for all recruits, black and white. One of the last drills we did in training was the church raise. The church raise drill is a fifty-foot ladder extended in the middle of the drill grounds, held by six recruits, with two holding the ladder poles and four holding a lifeline on four corners. In this drill, the recruit climbed to the top of a fifty-foot ladder, tied a ladder rope around his waist and leaned back with the arms straight out and the body perpendicular resembling a cross. This drill tests endurance and stamina. The training officers, Captain William Eubanks and Lieutenant E.B. Selph, appeared pleased with our progress. Their echoes of previous promises regarding our own black officers were grounds for encouragement. However, it would be fourteen years before one black firefighter was promoted and many years before we would command a company. Other actions had to be in place before this promise came a reality. Nonetheless, not keeping the promise created problems for the Memphis Fire Department. We kept the faith; they didn't keep the promise.

After lunch we sat in the fairgrounds under a tree, shaded from the hot sun. Captain Eubanks, our instructor, talked to us about our final grade averages. He didn't say anyone failed the test. However, he called out a few names and a few grades. The guys whose names were called were happy and shouted out when they learned their grades. Captain Eubanks called out my grade of 97. I said nothing. He continued, "Crawford got the highest grade-point average and he is showing no emotion." I was happy to finish training.

We learned later that our class was the first to be tested. Firefighters hired prior to our class, #16, were hired without qualifications such as a high school diploma. Compared to firefighters before us, we had class. Firefighters hired after the black pioneers were required to have a high school education. One or two of the white firefighters in the class behind us also had some college training.

Chapter 6

Unwritten Rules

The Memphis Fire Department, in certain instances, reflected the attitudes of the city. However, all white firefighters were not prejudiced or treated black firefighters unfairly. Most prejudiced black firefighters held their feelings in check because black firefighters had too much to lose. We had some good instructors who tried to be fair in grading performance and written work during training. Sometimes white officers were chastised for giving us fair grades that we truly earned. Most problems regarding fair grades in written and performance tests occurred later when we took promotional examinations. In time we experienced the pen being mightier than the sword. The only difference was we had to battle both: the pen that arbitrarily graded and the sword of ingrained prejudice.

Nineteen fifty-five was an eventful year. The U.S. Supreme Court banned racial segregation in public schools and the Interstate Commerce Commission banned segregation in interstate trains and buses. Many people flocked to buy color TV sets. The TV and radio stations played the Confederate song "The Yellow Rose of Texas." The other important event of September 1955 was that I was a member of the Memphis Fire Department and I had a good job that provided security for my family and me. One week following training we were assigned permanently to Station #8 at Mississippi and Crump Boulevard. We had A and B shifts. I was on A shift. Ours was the largest fire company in the city, since a company usually had six men and we had eight men on each shift. Each shift included six black firefighters, a captain and a lieutenant. Other stations had a total of only six men on each shift, including the captain and lieutenant. The captain and driver rode in the cab of the pumper. The rest of us rode on the pumper's tailboard. Can you imagine six men on the tailboard of a pumper? On the #8 pumper you could hardly get a hand on the bar and foot on the tailboard. We were lucky that we didn't lose a few of us before we got to a fire.

Basically, the lieutenant functioned as a captain at Station #8. Lieutenants at other stations shared the bed hall with privates. The lieutenant, as the white driver before him, shared the captain's quarters. His rank is a rank above a private's. He is the nozzle man; his duties were the same as a private, except at Station #8. Lieutenants at other stations stood watches. At Station #8 the lieutenant was excluded from watch duty and work. At all stations except #8 everyone was responsible for keeping his immediate area clean. At #8 we cleaned our area and we also cleaned the captain's and the lieutenant's area. They assigned us different tasks from those assigned white firefighters. We shined the captain and lieutenant's shoes. They'd say, "How about shining my shoes." They saw this as us doing them a favor and expected us to oblige. We made their beds, mopped the floors and cleaned their commodes and face bowls. My assignment was downstairs taking care of the pumper. I was responsible for apparatus maintenance although I was not officially promoted to a driver. I cleaned the pumper and dusted all furniture downstairs. Usually I was the only one downstairs. The captain and lieutenant ate breakfast and drank coffee while we worked. After I finished downstairs I'd go upstairs where I had breakfast with the other black firefighters. We never ate with the captain and the lieutenant. After breakfast we'd come downstairs and finish sweeping, mopping and washing tires on the apparatus. Sometimes we had training sessions in the mornings. During these sessions we studied together. During nights I studied on my own.

I felt more welcomed in Korea than I felt in the Memphis Fire Department. However, I think all of the black firefighters felt strength from the black community. Women in the community brought us dessert. In the afternoons old men sat on the bench in front of #8 and chatted with us. Black people would drive past, honk, wave and smile. I shall never forget the cheers the first time we passed through Mississippi and Walker on our way to a fire scene. Cheers rang out as I drove the fire truck through black neighborhoods with black firefighters on the tailboard of the pumper. I'd drive through Mississippi and Walker with the siren wailing full blast. Traffic in all directions stopped and surrendered the street to the fire truck. The loud blasting noise and red flashing lights from the truck brought people out of homes and businesses and commanded attention from people on the street. The fire engine's wailing siren continued until it stopped at the fire scene. I remained with the pumper. I observed little boys watching black firefighters, dreams of becoming firefighters in their eyes. I never had these dreams because I never saw firefighters that looked like me. I had no visible role model to anchor that dream. Our people were proud of us and they counted on us to make them proud. We had to succeed because we had reasons to be the best. I felt that as black firefighters all of us had to prove to others, and most importantly prove to ourselves, that we were the best. We had much to prove and many promises to keep.

The Memphis Fire Department hosted the Fire Department Instructor's Conference (FDIC) in Memphis every year. Firefighters came from all over the United States to Memphis for the FDIC. Black firefighters were right here in the city, part of its department, and they did not invite us to attend nor detail us to work overtime at the conference. Details to the conference included registration, driving the guests to and from hotels and demonstration sites, participating in the demonstration, dishwashing, pot washing and cooking. We complained to fire administration because we were not detailed to the conference. Administration ignored our request. We wanted to earn the overtime pay. Finally, they involved us in the conference. William Carter, Carl Stotts, Floyd Newsum, Elza Parson, Lawrence Yates and I worked the conference. A captain and a lieutenant were detailed with us. They were in charge, but they did no work. The black firefighters were assigned the pots-and-pans-washing detail, a detail nobody else wanted. The large pots looked like 55-gallon drums. I became so angry I picked up a heavy aluminum pot. Instead of walking the barrel pot to the other black firefighter, I flung it, sending it thundering across the concrete floor. Hearing the pot hit the floor; the captain and lieutenant came to see what was going on in the kitchen. From the look on their faces, I felt certain that I would be suspended for my behavior. Captain Maguire and Lieutenant Payne knew that I was angry. They knew we were all angry about this "shitty" detail. Undaunted, for the next few years, we continued kitchen duties. However, unusual events were about to occur.

The unusual events had officers moving quickly to avert what they saw as a potential embarrassment. It seemed that somehow a group of black firefighters were accidentally invited to the conference. I say accidentally because I am certain had they known the firefighters were black, they would not have been invited to this conference. They couldn't openly refuse black firefighters' admittance because people from all over the United States attended the Conference. However, the Memphis Fire Department certainly didn't expect black firefighters. Even though there were many minorities at the conference, accommodations were not set up for the black firefighters. Hotel, transportation and other arrangements were usually made by the host city. Fire administrators contacted fire headquarters fast and ordered a black firefighter to immediately report to the fire conference in class-A uniform. Class-A uniform is full-dress firefighter uniform. Well, now you couldn't have white firefighters escorting black firefighters during those times. Therefore, the department detailed Private Lawrence Yates to escort the black firefighters while they were in the city.

Private Yates, brawny, easygoing and good-humored, contacted us at the fire station. He explained the situation and the reason he was so abruptly detailed to the conference. Black firefighters met the uninvited guests at the Lorraine Hotel on our next day off. I was anxious to hear how they got invited to the conference. The black firefighters from Missouri explained that nothing was on the form regarding race. They said they simply completed the form and returned it to the Memphis Fire Department. We discussed the problems we had with our fire department.

The black firefighters from Missouri had no white firefighters in their department. We made jokes telling them that they attended a conference that we'd never been involved in. We agreed that they were there because administration was unaware they were black. We learned the firefighters were from several small counties in Missouri where the population was all black. We took them around the city. During the day they attended the conference as planned. At night we took them to the Flamingo, a nightclub off Beale on Hernando Street. The following years, administration automatically detailed us to the conference; they repeatedly assigned us kitchen duty.

Promotions for black firefighters did not come because we would never be "in" with those who ran the department. Nonetheless, the administration assigned Norvell Wallace on his shift and me on my shift to drive the pumper at Station #8. After Private Wallace was drafted into service, Private George Dumas was then assigned to driver for that shift. There was no test for drivers at Station #8. The assignment was a subjective evaluation by administration. It was based solely on what the officers saw and their opinions about the way you handled heavy equipment. They gave each of us the opportunity to drive the pumper. A few of the firefighters were not interested in becoming a driver. The rest of the men took turns driving the pumper. I felt comfortable driving the pumper. I had experience with heavy equipment as a truck driver, vehicle mechanic and dispatcher in the army.

My experience in service gave me an edge on those who had less experience driving heavy equipment. We role-played all positions during initial fire training. I made it a habit to know drivers from other companies. I fought fires vicariously since I had to stay with the fire truck. Communication between white drivers and me were friendly and with a sense of mutual respect. One of the drivers from the other company told me the captain, lieutenant and the training instructors' opinion was that I handled fire apparatus better than anyone.

George Dumas and I were drivers but not paid until March 1957. George Dumas was officially promoted to driver on B shift and I was promoted to driver on A shift. This was the first promotion since we were hired in 1955. The pumper driver was responsible for determining the most direct route to fire emergencies. This entailed knowing the streets, the location of fire hydrants, driveways and bridges and knowing the location of sprinkler connections. I spent hours learning the streets and businesses of South Memphis. I was familiar with some parts of the territory because I'd attended Catholic school in the area. However, I knew next to nothing compared to what I had to learn.

I spent my days off learning the territory. On some Sundays after mass at St. Augustine Church I'd take Delores and our two daughters, Joan and Jan, with me while I learned the territory. While riding through the streets and alleys of South Memphis, I thought of the bad weather and terrible roads that I had covered in Korea. There were few barriers in Korea. The job was done no matter what the conditions were. I worked in all kinds of weather, on all kinds of roads. Having to maneuver heavy equipment was nothing new to me. I looked forward to doing it again. Ironically, I found that some requirements of driving a fire truck were basically the same requirements as driving a truck in Korea. This included inspecting equipment for operation safety and maintenance of vehicles. At the change of shift each day, I checked the apparatus for damaged or missing tools. After each fire run I made sure the apparatus was working properly for the next run or response to a fire. I kept the equipment clean and shining so that when that engine rolled out of the house it looked brand new. We all took pride in the way our equipment looked, and I took pride in the way the equipment worked. Another important daily routine for me was studying fire department rules and regulations and other fire manuals.

Each time we answered a fire call, I realized the importance of my role as the driver. I maintained proper vision, knew the territory and got there quickly and safely. I prayed each time I climbed into that pumper and each time I returned safely to the station. While we fought our share of fires, we also answered our share of false alarms. Children, and sometimes adults, pulled the alarm and left the scene. I remember Station #8 receiving false alarms around ten o'clock every night. I observed, as I drove the pumper to one particular box, that we'd meet a city bus in the same vicinity each night. I mentioned my observations to Captain Don McGuire. Captain McGuire notified the other shift's captain. This captain also noticed the same pattern.

The frequency, the time the alarm came in, and the bus schedule were, I felt, more than chance occurrences. I figured the box-puller was riding the city bus.

Memphis Desoto Bridge (I-55). *Photograph courtesy of Robert Crawford.*

I devised a plan to catch and identify the box-puller. The on-duty fire marshal carried me to a bus stop approximately three miles from the firebox where the false alarm registered. I boarded the bus. Few passengers were on the bus. The bus arrived at the designated bus stop. Two people got off the bus at St Paul and Lauderdale Street. I also got off the bus at this stop. The firebox was on the opposite side of the street. One person—a teenager—and I went in opposite directions. I observed the teenager pull the fire alarm. After he pulled the alarm, the young man made an extremely loud sound mouthing that of a siren. He yelled wildly "arararararar, m—f—" Following his wild sounds mixed with profanity, the boy sat on the steps of a housing project apartment near the scene. He waited for the fire department. I held my place behind a bush across the street from where the young man waited for his stage show. Within minutes the show began. The drama began with the sound of sirens in a distance. Sirens blasted and red lights flashed. Two minutes later, firefighters were on the scene. The fire pumper, the ambulance and the fire marshal arrived simultaneously. After the fire department arrived, I left my post. By this time a small gathering of people watched the action. I pointed the young man out. The fire marshal transported the young man to juvenile court.

At the court hearing, the young man presented a letter from his principal and teacher. The principal and the teacher wrote letters regarding the boy's good character. He also presented a letter from his employer stating that he was a good employee. My testimony left no doubt the boy was guilty. The young man was placed on probation because of an excellent past record. The fire department's action and the court hearing negated more false alarms from this location.

Our company covered a large area of the city. Therefore, Station #8 was one of the busiest in the city. I knew every nook and cranny in this densely populated South Memphis area. Some houses looked like they may ignite from the heat of the sun. Some houses were empty while others were occupied. A section with attractive homes and manicured lawns was noticeably beautiful. The lawns showed no evidence of children killing the grass. Old houses passed from one generation to the other were well kept and they stood like castles with care and pride. On the other hand, our territory had more than its share of old dilapidated houses that stood like old fortresses, and many of them were firetraps.

Because we responded to so many fires, our hoses were always on the floor drying. A good hose was essential, and taking care of it was important. Everyone was responsible for washing the hose after we made a run. We used a mild solution of water and soap to remove grease and oil from the hose. The hose's cotton jacket absorbs oil and gas. The oil and gas deteriorates the cotton jacket and rubber lining. Since our station was too small to keep all the wet hose on the floor, we needed more space to lay the hose, so we took some of our hoses to Station #10. We had a hose cabinet at Station #8 that held twelve sections of hose but no hose tower. A hose tower is that part of the fire station or building designed so that the fire hose can hang vertically to drain and dry.

After working with the hose you had a pretty good thirst. A cup was on the water fountain for blacks in the neighborhood who passed by and wanted a drink of water. We did not use the cup. To deter us from drinking from the water fountain, white firefighters attached an "out of order" sign on the fountain. Sometimes they turned the water fountain off. We complained to the chief about the water fountain. Instead of making them turn the water back on, the chief stopped us from going to that station to hang hose. We had the same problem at another fire station.

One advantage of different companies making fire runs together was that regardless of race, you became acquainted with each other, for better or worse. Another advantage included white firefighters observing black firefighters in action. They could not deny that we were excellent firefighters. Pumper drivers and I became acquainted. It is amazing how differently people act when they are one on one rather than in a group. Joe Halmer frequently made runs with us. Joe was ruddy with work-swollen banged-up hands. One of our jobs on fire scenes was to make sure the men got proper water pressure to fight fires.

Sometimes a driver from another company and I parked close and hooked fire hoses to each other's pumper. This depended on who was closer to the fire hydrant. We helped each other with fire hoses and we helped each other get water to the fire. I got news about what was going on in the department from drivers. Officers at Station #8 told us nothing about what went on in the department. We had no idea what was happening in the department. I felt one of the first things you ought to do is learn something about the organization of which you are a part.

"How long have you been on," I asked.

"Seven years tomorrow," Joe responded. "Been driving the pumper about two years."

"I saw you at the fire on Dunnavant," I said. " Fortunately, that fire was mostly smoke from burnt neck bones left on the stove."

"We cover for #12. Y'all get lot of fires at #8, don't you?"

"Yeah. Lot of false alarms and burning garbage dumpsters in the projects. But man, when we have a big one, it's really big. Get a lot firefighting experience at #8," I answered.

"Not when you're driver. You just watch the truck while others get the action," Joe said.

"I know. I know I gotta have firefighting experience to do what I want to do on this job."

Private Carter saw me talking to the white driver and asked, "What you talking to him 'bout?"

"I'm getting information," I said.

"What information?"

I turned to Private Carter and said, "I'll tell you later."

Because we were isolated at Station #8, talking to white firefighters was one way of learning how the department operated. I remember my father admonishing his congregation. He said, referring to preparation for heaven, "the time you start getting old and sick is not the time to start preparing." I gathered information about the fire department every chance the opportunity presented itself.

Although black firefighters functioned as a group in job-related issues, we maintained our individuality. There was room for each of us in the struggle and each contributed in his own way. All of us kept our eyes on the problems. Since knowledge is power, we were limited in the sense that we were not privy to knowledge about everyday workings in the department. Talking to white firefighters was one method of learning every day operations of the department.

Since talking to white firefighters at fires was one way of getting information, I made it one of my priorities. Most of the time there was a grain of truth on the grapevine. I felt it wise to get to know as much as possible from all sources about the fire department. In many situations discussing the fire department with white firefighters was beneficial. For example, one of my tasks as pumper driver was to complete monthly reports. A couple of black firefighters chided me for completing the reports because they felt this was the captain's job. One black firefighter said to me, "You ought not to be doing his work." Another one responded, "That's his responsibility. Let him do his own work."

I learned from talking to other pumper drivers on fire scenes that it was common practice for drivers to complete monthly forms. Since most of the black and white firefighters did not communicate, black firefighters did not know it was common practice for drivers to complete monthly reports. Nonetheless, I did not worry about doing the "captain's work." I was willing to do more than I got paid for as an investment in my career as a firefighter.

I drove the pumper for thirteen years. Having to remain with the pumper afforded me no opportunity to fight fires. I knew that I needed firefighting experience before being promoted to lieutenant in the department. I observed that most of the men had experienced each level as they went

Robert with fellow firefighters, Privates Martiniano Lerma, Warren Uselton Jr., driver Carl Johnson, Captain Paul Masnica, Lieutenant Robert Crawford and Private Gary Green. *Photograph courtesy of Don Lancaster.*

through the ranks. I figured the aerial ladder truck would be the next step for me to take to get firefighting experience.

In 1957, while the new Claude A. Armour Training Center was taking shape at Flicker and Avery in Memphis, Tennessee, in Little Rock, Arkansas, racial violence prompted President Eisenhower to send a force of some one thousand U.S. Army paratroopers to enforce the desegregation of Central High School. The nine black students entered the guarded school on September 25.

At Station #8 we never discussed segregation issues in the engine house. It was as if we had a code of silence regarding integration with white firefighters. Even though no one discussed it openly, segregation issues were on our minds. Racial problems during 1957 dominated TV. I felt I needed to be preparing for changes that were about to take place in other areas of the workplace. No doubt that things were beginning to take a turn. The schools were the starting point. I thought about how brave the nine black children were to face an angry mob of whites and enter that school.

In 1966 the movement for school integration showed signs of division, as did the entire civil rights movement. Followers of Dr. Martin Luther King Jr. differed with militant Stokely Carmichael. The terms "black power" and "white backlash," the corollary catch phrase, were coined. James Meredith was shot and wounded while on a lone march from Memphis, Tennessee, to Jackson, Mississippi, to encourage black voter registration.

Some changes were also occurring in the Memphis fire department. There had always been one-sided integration in that, of necessity, we had a white captain and lieutenant at Station #8. Black firefighters were never housed at stations other than #8 until the fire department integrated its engine houses in 1966. All black firefighters except the two drivers, George Dumas and I, transferred to other stations. White firefighters designated one special bed that they assigned the black fighter. I called this the "black bed" because this was the only bed that a black firefighter could use in the bed hall. Black firefighters on each shift slept in this "special" bed only. This was the pattern all over the city where black firefighters were stationed.

Chapter 7

Not Without a Fight

It was spring 1968, a time of unrest for the city of Memphis. Sanitation workers were on strike. Violence erupted during a march led by Dr. Martin Luther King Jr. in support of the sanitation worker's strike. A black marcher was killed. Dr. King urged calm as National Guard troops were called to restore order. Dr. King promised to return in April to attend another march.

Two black firefighters, Privates Norvell Wallace and Floyd Newsum, were stationed on opposite shifts at Fire Station #2. Private Wallace was brown-skinned and stocky with a great sense of humor, but he had a no-nonsense attitude toward matters of importance. He was active in the air force reserve. Private Floyd Newsum was a civil rights activist in the sanitation workers' strike. Station #2 was directly south of the building where the assassin fired the shot that killed Dr. King. Fire administration transferred the two black firefighters from Station #2 on South Main the night before Dr. King's assassination at the Lorraine Motel.

On April 4, 1968, the country was shocked by the assassination of Dr. Martin Luther King Jr. A week of rioting in black neighborhoods followed. I drove the pumper during the strike. We worked our regular twenty-four-hour shifts at Station #8. After leaving our station we went to Station #10 to put extra equipment in service because there were so many fires in the black neighborhoods in South Memphis. White firefighters had their own guns on the apparatus.

Administration learned that white firefighters carried firearms on the apparatus. The firefighters were ordered to remove the weapons from the equipment. Police were dispatched if we needed them. I am thankful that we didn't have to break up demonstrations. That could have spelled trouble

Housed in the Historic Lorraine Motel where Dr. Martin Luther King was assassinated, the National Civil Rights Museum is the first museum of American civil rights. *Photograph courtesy of Robert Crawford.*

for firefighters as well as the demonstrators. We answered one fire call after another all night. We handled all types of fires during the strike, including fires in vacant buildings, cars, and garbage dumpsters in the Foote Homes Housing Project. People stretched rope from telephone post to telephone post. Furniture blocked Lauderdale Street to keep the fire department out or to slow down our response time.

Private Newsum resigned from the department after administration denied him a leave of absence to deal with personal matters resulting from the strike and Dr. King's assassination. Meanwhile, turmoil continued into the fall of 1968. Dissident students went on strike at San Francisco State College, calling for reforms, especially in black studies programs. During almost four months of turmoil, the president of the school called in police to help end the strike. In our station house, TV aired the situation in San Francisco, but I was not certain that Captain Maguire was watching with interest.

It was mid-afternoon. Most of the firefighters were upstairs napping in the bed hall. I finished reading about aerial truck operations and duties of the driver. I started tying the rescue knot. As I worked on tying knots, I knew that I had to turn in a request-for-transfer form and go through the chain of command in requesting a transfer. Therefore, I would begin first with my immediate supervisor, Captain Don Maguire, and then second supervisor, District Chief Herschel Dove. Finally I would to talk to the chief of the fire department, Chief Eddie Hamilton. So I began my tasks. I sat facing my immediate supervisor.

"Captain Maguire," I began, "I'd like to transfer to Truck Company #5." He looked puzzled. I continued, "I need experience in fighting fires. As ladder truck driver I can get the firefighting experience I need to become an officer in the future."

"I'll have talk to District Chief Dove," he said between sips of coffee. "I'll let you know what he says."

I knew that it would take a few days before I would have an answer because District Chief Dove had to talk to his supervisor. Chief Dove became supportive later in my firefighting career. Captain Maguire later informed me that my request for transfer to Truck Company #5 was denied because I was not qualified as an aerial truck driver.

"How do I become qualified to drive the ladder truck?" I asked.

"You have to be on a company and train on that company."

"Would you make an appointment for me to see Chief Eddie Hamilton concerning the transfer to Station #2 so that I can drive the ladder truck?" I asked Captain Maguire. I was confident that I could drive the ladder truck. All I wanted was the opportunity. I could do the rest. I was fully aware that I would be the only black firefighter on that shift. Going back to Korea would be easier than working with some of these attitudes, I thought to myself. I never promised myself it would be easy. But this was the only way that I could place myself in the position for promotion to lieutenant. I am thankful that I learned the art of networking in the army, Orange Mound Civic Club, church and other organizations with which I had been involved. My long-range plans included being prepared to take any exams offered by the department and to always have a passing grade for promotions that would become available.

District Chief Dove made an appointment for me to meet with Chief Hamilton. The appointment was on my day off. I arrived at Fire Headquarters ten minutes earlier than my appointment. The secretary offered me a seat. Soon Chief Hamilton invited me into his office. The chief was five feet ten inches tall and of medium build. He was known to tear ass. I had no idea that I would experience his wrath in my future. However, on this day I found him easy to talk to, and I was not the least intimidated in his presence. We talked shop for a while about the department. This gave me an easy lead-in to explain my request. Chief Hamilton sipped his coffee, set the cup down, looked directly at me and prepared to listen to what I had to say.

I explained to the chief why I wanted to be transferred to the aerial truck.

"I want to be promoted to lieutenant in the future," I said. "Since I've been with the fire department, I have been a pumper driver. I don't have firefighting experience. I will be more effective as an officer if I have firefighting experience," I explained.

"What kind of lieutenant grade do you have?" he asked.

"An 85," I answered.

"See if you can get a higher grade."

"I can do that, "I said with certainty. I thanked the Chief. I left his office with the determination to improve my test score on the next exam. The next exam would be given the fall of 1968.

I concentrated intensively on studying all the fire department's manuals. I meticulously observed the difference in policy and procedure according to the manual and the way some officers carried out or neglected to follow procedures. I learned long ago that climbing the ladder one step at a time is best. The goal of reaching the top, I think, is knowledge of how we form the bottom. I learned the qualification for a ladder truck driver. Most importantly, for me, is that the aerial truck driver fights fires.

Arriving to work the following day, Captain Maguire informed me that I had orders from Chief Hamilton's office to report to the fire academy to qualify for aerial truck driver. I reported to Armour Training Academy at 8:00 a.m. This was a good day for me. On my way to the academy, I thought of everything I learned about maneuvering equipment in service. I reflected on what I learned from truck drivers in other companies. Most importantly, I visualized what I studied in firefighting manuals and what I observed watching firefighters in action. I was confident that I would qualify to drive the aerial ladder truck. I had driven the pumper for thirteen years. I entertained no doubt about my qualifications.

I arrived at the training academy at 7:45 a.m. It was a clear Memphis day and the sun was warm. I scanned the grounds for the aerial truck that I would drive. My eyes focused on the large, red one-hundred-foot Pirsch aerial truck with a gas engine. The equipment sat almost in the center of the drill field. I was anxious to drive the apparatus.

Chief Selph, chief of training, had a professional demeanor. We climbed into the truck, I took my place under the wheel and he instructed me to position the truck for different aerial rises. Chief Selph observed how I operated the truck. I drove around the drill ground. I parked the truck; I spotted or put it in position for aerial operation in places where Chief Selph directed.

Three, then four hours passed. I continued operating the 100-foot aerial truck. I drove the aerial truck around the training center half the day. Maneuvering here and there, backing up, turning around, raising and lowering the ladder and parking the aerial truck at various angles. Chief Selph watched with military precision. He said very little. I felt totally in charge. I manipulated the ladder truck extremely well. Chief Selph concluded that I handled the tractor quite well but he felt that I couldn't handle the tiller wheel, the rear driver, on the 100-foot aerial. Chief Hamilton was only concerned with how I handled the front end of the ladder truck. After I'd finish driving, Chief Selph said, "Yes, you can handle the tractor part ok. You are qualified to drive the tractor."

The next workday, Chief Hamilton transferred me to drive ladder truck #5 at Station #2. I was where I wanted to be at that time, in a position to get firefighting experience. I had no idea that my ambition to drive the ladder truck and my desire to fight fire would later be used against me.

Following integration of fire stations in 1966, one black firefighter was assigned to each shift at each designated station. I was aware that I would be the only black firefighter on my shift. Norvell Wallace and I slept in the "black bed" on our respective shifts. Fire Station #2 housed three companies—the pumper company, the truck company and the salvage company. Captain Paul Waddell was transferred to truck #5 three months after my transfer to that station. He was a young captain, the same age as me and with less seniority on the job. He was about five feet nine inches tall, moonfaced and seemed to prefer his own company to others. He seemed to like no one, including white firefighters. I knew where I stood with this guy. Consequently, I found his remarks agitating. For example, we watched football games. When a black player ran the ball, the captain yelled, "Look at that monkey run."

I remember a specific day when Captain Waddell felt like picking a fight. Naturally I was his target. He said with bitter sarcasm, "One of these days we will have 'Nero' for fire director and then we can tap dance and play a fiddle while the city burns." I was ready to punch the captain out that day. At that point I remembered what Carl Stotts, a Pioneer, told us. One of the white firefighters said, "We don't want you and we will do our best to run you away." I felt this was Captain Waddell's motive. I knew the captain was being provocative. Everyone knew I was angry. I could feel veins in my neck standing out in livid ridges. I knew others saw what I felt. Private J.C. Martin was friendly and the quietest man at the station. He rarely spoke to anyone. He came to me and said, "Don't do it; we all know." Jerry Wright, another firefighter agreed. It was a very trying time for me. I was aware that striking an officer was automatic grounds for dismissal. I came so close to losing it that day.

Captain Waddell continued bitching and nitpicking with me for a long time. I didn't feel bad about my intense dislike for this man because this captain was not popular with anyone on his company. He constantly nitpicked about every thing I did. For example, as truck driver I had the same responsibility as the pumper driver. I did all checks at change of shifts and after each run. Captain Waddell said fingerprints were on the door handle and hood latch of the aerial truck. He suspended me for that. He also said that I left a bucket of water in the middle of the floor after we finished cleaning the truck. There may have been fingerprints, but someone else left the bucket of water there. But he said I was in charge of the work and I should have told someone to move the bucket of water. Captain Waddell assumed a posture of superiority and said, "You're suspended. You'll be notified when to appear for your hearing with the chief."

I stared at him momentarily and then I promptly left the premises. As I left the station I had a feeling Captain Waddell knew I had taken the lieutenant's test. He was aware that I had made a good passing grade. It was the competition that he couldn't deal with and not the fingerprints or any of those little things that bothered him. There was no doubt in my mind that he was trying to block my pending promotion to lieutenant. I was becoming aware that I was a threat to the "good old boy network" because I asked questions, consistently read manuals and I was not afraid to approach the director himself. I was my own network. Needless to say, my attitude did not sit too well with Captain Waddell. The lies that black firefighters were afraid of fire and therefore could not fight fire stayed with us for a long time. This lie almost cost me the most important position on the fire department.

I am thankful that my past experiences included decent white people before meeting some of the white guys with the fire department. I knew that they were no more all alike than blacks. Thank goodness that life taught me early on the importance of recognizing, understanding and accepting individual differences in people. Nevertheless, I was informed that I was to appear at fire headquarters for a formal hearing with Chief Larry Williams.

At the hearing, Captain Waddell brought up issues that had nothing to do with why he suspended me. In fact the issues he raised surprised me. I felt it was one of those "I got you where I want you now" games people play.

"Crawford admitted that he couldn't fight fire. He said himself that he wanted to drive the truck so he could learn to fight fire," Captain Waddell said to Chief Williams.

In spite of the lie regarding black firefighters being afraid and having poor firefighting skills, some white firefighters respected us as good firefighters. Sometimes a deputy chief would commend us on our firefighting skills. They knew we were good firefighters. Some of them trained under us at Fire Station #8. Captain Waddell also said that he had to tell me where to place the ladder on a building during a fire. He also said that I damaged a lawnmower. Of course, he explained that I left fingerprints on the hood and on the door handle of the truck. Some of his charges were news to me. The man dug deep in his bag for every minor infraction and made it seem like a great catastrophe. However, I flinched when he attacked my ability to fight fire. Captain Waddell knew I was quite capable of handling a fire.

Criticizing a firefighter's firefighting skills is the worst accusation one firefighter can make against another firefighter. He repeatedly emphasized that I could not fight fire. He accentuated the fact that I admitted I could not fight fire and that was the reason I wanted to drive the ladder truck. After hearing complaints about all of my "inadequacies," I realized Captain Waddell would have the last word and I was prepared to take my suspension or worse. However, I was surprised at Chief Williams's decision.

Chief Larry Williams steeped his fingers like praying hands and assumed a judicial expression. "Well now," Chief Williams said, "he couldn't be all that bad. He's been on the job fourteen years. I know he learned something during that time."

I appreciated the chief's looking at the determining facts in this situation, rather than looking at who was determining the facts. Chief Williams did not uphold the captain's recommendation. Captain Waddell fumed at the chief's decision, gathered his belongings and stormed out of the hearing. I returned to work the following day.

Captain Waddell always watched me, but after the hearing he watched me with the vigilance of a hawk. For example, we were fighting a fire and instead of the captain watching where I placed the ladder, he gave me a predatory expression. I am not certain if the captain's face was hot from the heat of the fire, but his face appeared hot and pinched with resentment. He watched eagerly for me to make a mistake. If I made a mistake I felt it would probably be the end of my career as a firefighter. I knew I was under the tough scrutiny of this captain. Therefore, I made certain that I used firefighting procedures. But I didn't worry about Captain Waddell, as I was concerned about knowing the job and doing it well. I was confident of my abilities. I knew how to fight fire and I didn't learn from Captain Waddell.

I could not allow imprecise details to get in my way. I had to be certain of every detail. I was not really as concerned about impressing Captain Waddell as I was about impressing myself. I did not expect the captain to be positive about my work. In a way this was good; it helped me become a better firefighter. I could rest assured that Captain Waddell would certainly point out any mistake on my part, but I refused to let anyone make me feel inferior. I remained under Captain Waddell's tough scrutiny. He said little to me at the station, and I avoided him. Of course, I conversed with him on matters relating to the job. He was distant during these times, but I had to do what I had to do. Since I worked under direct supervision of Captain Waddell, I was compelled to discuss job-related matters. Most of the time he'd clam up. I left him with his own thoughts. I did what I had to do as driver. I filled out all driver-related reports. In spite of all the problems Captain Waddell sometimes created for the whole engine house—and most often, me—I never fully understood his behavior. He did something that I must admit threw me completely off guard.

I was accustomed to working two jobs. Firefighters worked in twenty-four-hour shifts and then were off twenty-four hours. During my three days off, I had an accident while working my second job. I was repairing a pressure cooker. It all happened fast. The thing exploded. Hot steam gushed out of the cooker. I was severely burned on the upper part of my body. I couldn't imagine being in all the fires I have been in and being burned by a damned pressure cooker.

I was hospitalized for three weeks. I had just finished force-feeding myself that terrible hospital food. I pushed the tray away after surrendering to self-inflicted torture. I looked up. I was certain that my heart was in good shape; otherwise I surely may have suffered cardiac arrest. I could hardly believe my eyes. Standing in the door was Captain Waddell. He even brought me a box of Philly Sheroot, my favorite cigars. I am not a vengeful man. I accepted the olive branch. We carried on a civil conversation and chitchatted about the fire department, the aerial truck and other mundane things. His cordial manner was a side I'd never seen. Although his visit was short, it was pleasant. None of the other guys at the station visited me during my hospital stay. Regardless of any guilty feelings this man may have had when he visited me that day, his anger had already done damage. It would be many years from now when this same lie that I could not fight fire would almost cost me an important promotion with the City of Memphis Fire Department.

I left the hospital and returned to work within three months. Captain Waddell no longer nitpicked. He remained reserved when I was around, but it was a more acceptable reserve.

Chapter 8

Separate But Not Equal

Black firefighters at Station #8 were limited in many ways. We inspected no buildings without a white officer. White privates had the privilege of inspection without an escort. We were not given detail to other stations. They confined us to Fire Station #8. We had second choice of vacation and days off. We were not allowed to participate in the department's instructor conference. It was as if we were not part of the Memphis Fire Department, but were an entity unto ourselves. Isolation and our supervisors withholding information were problems for us.

Our captain and lieutenant didn't keep us abreast of what went on in the department. My alternative was to use my time wisely. I read about new techniques, equipment and procedures and I observed others at work. I read *Fire Engineering Magazine* and other firefighting magazines faithfully. I listened to the grapevine and asked lots of questions about the department. I talked to white firefighters about the department. I knew that we were pretty much on our own. Sometimes we were unaware of new equipment. For example, when I was a driver at Station #8 I remember the time we answered a fire call to McLemore and College Street. When we arrived, other fire companies were already on the scene. Thick, heavy smoke billowed out of the structure. We went into the burning house and knocked the fire down. White firefighters with breathing apparatus and some black firefighters without breathing apparatus were passed out. Chief Doyle arrived on the scene. Chief Doyle was tall and slender with a receding hairline. I viewed him as indifferent to black firefighters and strictly fire-department-business oriented. "Where are your masks?" he asked apathetically.

"Masks?" two black firefighters asked in unison, "What masks?"

This was the first we heard about breathing apparatus. We were certain that we had not been issued masks. Some of the men on #8 were still passed out. Since no black ambulance was on the scene, Chief Doyle ordered a white ambulance company to transport black firefighters for hospital observation. That evening chief Doyle ordered breathing apparatus for the firefighters at Station #8.

We serviced areas heavily populated by blacks and we responded to all second-alarm fires in South Memphis. However, we did not operate the same as other companies. When we responded to a fire that the captain suspected was a white person's home, the captain instructed us to remain outside until he ordered us to enter the residence. He made certain white women were appropriately dressed before we entered the home. On the other hand, when we made a fire call to a black family's home, we entered and investigated the situation with the white officer. Because the same situations were handled differently, I labeled fires in black homes "black fire."

Sometimes our fire scene was more than fighting fires. For example, my shift worked the night of the incident that created great problems for us. I remember the issue that created an argument between a white fireman and black firemen. The situation at that fire became extremely heated, not so much from the fire as from human emotions. I was concerned black and white firefighters would end up fighting each other. This incident occurred following a box alarm at St. Augustine Catholic Church at 903 Walker Avenue. Many false alarms came from children pulling alarm boxes that were mounted on utility posts. The area was densely populated with many substandard houses that were essentially firetraps. We answered more calls than any station in the city. Lieutenant Nevels—a white officer, red-faced, tall and slender with a receding hairline—exercised poor judgment. At times our coping mechanisms fail because some situations defy acceptable labeling. Perhaps this one was one of those situations.

Since I was the pumper driver, I remained with the pumper. Lieutenant Chandler was at his pumper and I heard his radio conversation. In his report to the alarm office he stated, "A white man said he saw two nigger boys pull the box and run." The white man he referred to was my priest, Father Miro Wise. I did not hear the priest say this and I doubt very seriously that he would have used the term. Even if he had, the lieutenant would not have had to repeat the slur. I felt that Private Newsum would have been expressive but diplomatic in his approach. I planned to handle the situation as tactfully as possible. However, Private Carter, his face set in a permanent frown, was very vocal. He immediately let loose a barrage of unsavory choice words at Lieutenant Chandler. Needless to say, the firefighter's remarks became a hot issue. We were ordered to see Chief Hamilton at fire headquarters the next workday.

I drove the pumper to headquarters. The three of us were all nervous. We had made a mistake in the eyes of the department. We expressed our nervousness in different ways. Private Newsum was quiet, Private Carter's face grew pensive and I nervously bit the inside of my jaw. The three of us sat on a couch facing the chief in his office. The desk lamp obscured Chief Hamilton's full view and he could not see me. He turned a cold eye on me. "Move over so I can see you," he growled. The chief was blunt and vociferous in his reprimands. He made piercing, impolite eye contact as the edge of impatience crept in his voice. He glared warnings from one to the other of us. His hard, cold, pinched facial expression was worse than the wrath of any army sergeant. The chief dished out our penalties one by one. He charged Private Newsum with leaving his post. He chewed me out for leaving the pumper and also charged me with leaving my post. He charged Private Carter with insubordination and terminated him. Chief Hamilton also gave Private Newsum and me a ten-day suspension without pay at the convenience of the fire department. Since the department was segregated and since Private Newsum and I were on the same shift we could not be off duty at the same time, as this would make the company one man short. Both of us being off at the same time would entail detailing a white firefighter to Station #8. He would have to sleep in the bed hall with black firefighters. Consequently, it was expedient for Chief Hamilton to suspend us at the "convenience of the fire department." Lieutenant Chandler did not appear nor was he reprimanded for his part in the incident.

Lieutenant Chandler later apologized for the racial slur. The Civil Service Commission reinstated Private Carter. However, he did not return to work because Chief Hamilton appealed the Civil Service Commission's decision. Chief Hamilton stated that Private Carter would never return as long as he was chief of the Memphis Fire Department. This incident is one of many dilemmas that happened to pioneer black firefighters in our city. Incidents and unfair treatment such as this spoke poorly, especially for a fire department that was considered one of the nation's finest. Injustice and other grievances compelled us to seek fair treatment from the fire department through the courts. Changes would have to occur before Private Carter resumed his place as a firefighter with the City of Memphis Fire Department.

Even vacation time was unfairly scheduled. All vacations were from March to August. The lieutenant and captain always picked the same months, June, July or August. That left us with the other four months. We pulled straws. I remember pulling a March vacation, which is still a cold

Demolition of the first Fire Station #8. *Photograph courtesy of Robert Crawford.*

month. "What's my chance of getting a summer month vacation?" I asked Chief R.F Doyle.

"You have two chances," he answered, "slim and none". It was a snowy March that year when I had my vacation.

Station #8 had more than its share of fire calls because there were lots of fires in the area and because children often pulled fireboxes. Young white officers were sent to our station. It was well known that if you wanted a man to get firefighting experience, send the man to Station #8. We fought fires from 7:00 a.m. when we got to work until 7:00 the next morning. We trained young white lieutenants who had less experience and seniority on the job. Sometimes they became our captains and chiefs. Often these new officers had less seniority on the job. Some white firefighters did not mind coming to #8 because this was an easy assignment. They didn't have to work or do chores at our station because black firefighters did all the work at Station #8. However, they knew we were good firefighters. Some white firefighters accepted the fact that we were good firefighters and respected our firefighting skills.

Nevertheless, I was interested in holding the fire department to "the promise." Therefore, I planned to always be prepared for any test and to have a passing grade for any position the department offered. Some black

New Fire Station #8. *Photograph courtesy of Robert Crawford.*

firefighters became disillusioned and seemed to give up efforts at making promotions. Others had no intentions of rolling over. For example, I was studying one morning after we had finished chores. John Cooper, a tall, slender brown-skinned Pioneer, liked to make jokes. He came to where I was deep in the books and looked over my shoulder. I turned around to see what he wanted.

"I don't know why you waste your time studying that book. They ain't gon' never give us no promotion."

I looked at him. "Well," I said, "in case they do, they can't say I don't have a grade up or that I am not qualified."

"Still say you wasting time." He walked away and started teasing about something else.

I looked at John and thought, "I need these long-range goals to help me get past these short-term obstacles. I never intend to apologize for working." I resumed my study.

My goal on this job was to achieve the highest rank possible in the fire department. We had been given "the promise" and there were things I had to do to reach the goal. I liked the challenge. I have always been observant. I made mental notes of the captains' and lieutenants' decisions regarding the job. I read systematically, specifically concentrating on learning rules and regulations in the fire department's manual. A few of us decided that we would always be prepared for any test the department offered. I checked and compared the officers' decisions with rules and regulations according to the manual. I role-played the way I would have handled the same problem. On many occasions I observed poor judgment and apparent lack of leadership on the part of officers. I observed black firefighters being treated as though we were invisible. I observed some white firefighters being ignored if they were not in the "network." I saw rewards and recognition given to firefighters who obviously did inferior work. According to what I read in the manual, in many instances, good performance on the job was given little or no value.

Chapter 9
Gentlemen's Agreement

Promises regarding promotions and hiring black firefighters continued. We repeatedly observed young white firefighters with less experience and seniority promoted while we were told, "wait until next time" or "we'll get you next time."

Though Chief Eddie Hamilton made the first move and promoted two black lieutenants, promotions for black firefighters were moving slowly. A few of us seemed to give up the idea of being promoted in the department since we were not in the "good old boy" network. Some of us continued to study and to take exams each time exams were offered. However, even though we passed exams and had qualifying grades, we were consistently given excuses as to why we were not promoted to higher ranks. In the administrator's eyes we were never qualified. It was the same old story; we did the training and they got the promotions. New lieutenants that we trained repeatedly became our captains as result of the network. We had no one to speak for us. We certainly weren't going to make rank via the network. According to that system, it appeared that we would never qualify for promotion. If we scored high on the written exam, we never, according to our officers, scored high enough on the performance exam. The performance exam was particularly subjective. It included whatever the officers observed about your work or wanted to grade you on during the course of a year. I remember two white chiefs who were fair in grading us above average on the performance exam. Administration felt the white chiefs were becoming too friendly with the black firefighters. Because they gave us grades that we truly earned, they were disciplined by means of a transfer from our district to another district.

In 1965, Memphis Fire Department was cited nationally as having a Class One fire department. This same year, the City hired ninety white firefighters and only three black firefighters. The black firefighters were Lawrence Lum, Maurice Logan and Thomas Bethany. Ten years passed before the City found it necessary to hire more black firefighters. Hiring proved disproportionate in favor of white firefighters. The promises made regarding our own station and our own company seemed to fade into oblivion. Everyone appeared to be in denial about "the promise" except black firefighters. Administration's tactics disgusted me. I resolved to let it lead to change. We had a purpose. We would never see the promise fulfilled. We needed an organization. We needed a black firefighter organization, a voice to put teeth in our demands.

I remember two events that occurred in Memphis in 1966. The Beatles played concerts at the Mid-South Coliseum. Protesters, several of them in Ku Klux Klan robes, picketed outside because of John Lennon's remark claiming that the band had become more popular than Jesus. This same year William C. Carter was making an attempt to return to his job on the fire department.

Chief Hamilton dismissed Private William C. Carter from the Memphis Fire Department in 1964 on charges of insubordination, unbecoming conduct and using profanity. In 1966 the Civil Service Commission ruled that Carter should be reinstated. Chancellor Charles A. Ron upheld the Civil Service Commission's ruling. However, Commissioner Armour felt that reinstating Carter would be a "disservice to the Memphis Fire Department." It would take eight years from this date and the development of a strong organization before Carter and three other black firefighters would resume their places on the City of Memphis Fire Department.

In December 1968, following Dr. Martin Luther King's assassination, Chief Eddie Hamilton promoted Norvell Wallace to lieutenant. This was the first promotion of a black firefighter in thirteen years. It was also the first command position for a black firefighter since our hiring in 1955. Lieutenant Wallace was not assigned to #2 Engine House. He was transferred to truck 10 at #24 Engine House.

Most black firefighters made certain that they had a grade in case the opportunity came for a promotion. I started studying for a higher grade on the lieutenant's exam immediately following my transfer to aerial truck #5 as driver in 1968. I was doing what I wanted to do; I was gaining more experience in fighting fires. Once we got on the scene, I fought fire. However, I continued to study.

Tests were given once a year. Test grades were good for two years. I always had a passing grade up for lieutenant. In the fall of 1969 the test scores came in. My grade was 97.5. I was number five on the eligible register for the position of firefighting division lieutenant. Usually Chief Hamilton made promotions during Christmas holidays. Promotions during this time were his Christmas present to the men. In December 1969, I was promoted to lieutenant, effective January 1, 1970. I had worked for the department fourteen years before I made this promotion. I was transferred from Station #2 to Station #6 at 924 Thomas Street. I remained in firefighting. I was where I wanted to be at this time because I got more experience in fighting fires. After my promotion, Captain Waddell and I never worked again on the same company.

Plans for integrated fire stations started when we were at Station #8. The NAACP and black leaders were concerned about situations at the fire stations. The issues revolved around hiring and promoting firefighters and integrating the fire stations. Since the stations were now integrated, Private Nathaniel Partee was the only black firefighter at Station #6. He was transferred to another company when I went to that station. It appeared the plan was to have only one black firefighter on each shift at fire stations. This was the pattern all over the city. Apparently it was easier for whites to deal with one black firefighter instead of having to deal with two or more on the same shift. It seemed that black firefighters were not as intimidated in the company of white firefighters. On the other hand, most white firefighters appeared uncomfortable around black firefighters. It appeared that the more blacks, the more discomforting the experience.

When I transferred as lieutenant to Station #6 at Thomas and Chelsea Streets, the area around the station was in transition. Most of the white families and businesses had already moved out of the area. Three companies were in that engine house. There were three lieutenants and three captains. I was lieutenant on the salvage corps. My primary function was saving furniture, equipment and anything in a building during and after a fire. If contents of a building or residence was salvageable property, I put waterproof covers over the contents or moved the contents outside to provide protection. I placed temporary covers on roofs and other openings to protect the interior of the building or residence. If nothing was salvageable, I joined in and helped fight the fires. I felt capable working in any fire department operation. However, some officers would choose to keep me limited to salvage corps only.

The three captains at the station made no denial regarding detailing me. I would take charge when Captain Armstrong was on a day off. However, I was never detailed outside of Station #6. Detail means taking charge of another officer's company when that officer was off duty. They had their little innuendos regarding detailing me. I felt they played a game of tic-tac-toe with their vacation and days off. They seemed to go to any extreme to prevent me being in charge of their companies. For example, the captains

Pyramid Arena. This river landmark is used for exhibits and concerts. *Photograph courtesy of Robert Crawford.*

made arrangements where they would cover for each other so that I would not be in charge of their company. They detailed the other two lieutenants. These lieutenants were detailed and permitted to operate on their own. If you worked under the captain, that meant you were not in charge. It was their way of trying to show me I was incapable of making good fire ground decisions. I did not accept these officers trying to determine my worth. It was apparent they did not want a black officer in a position of authority, making decisions, or being in charge of an all-white company. For example, when an officer was absent from work, an officer from another company was sent to take charge, even though I was in the engine house. They knew that I could have easily filled in for the absent officer, but they made certain that I remained under the watchful eye of my captain. Nonetheless, I concentrated on my long-range goals. I knew things that came too easily could never enrich me more than adversity. I was patiently angry.

I felt it was not so much what the captains were doing. It was what I chose to do about what was happening. What I chose to do about the situation would make the difference in how my career as a firefighter turned out. Since I was not detailed outside of #6, and since it was the officers' individual decisions, I decided to take care of the matter. I was aware that personally I was not a threat to some officers; the real threat was the color of my skin. I learned the responsibility of a firefighting lieutenant from reading the manuals. Not being detailed to other stations was limiting me. I chose to challenge the captains' decision regarding details. I refused to leave my future in the hands of these captains.

I was fully aware that chiefs and deputy chiefs made decisions at headquarters. Therefore, I risked going too far to discover how far I could go. I did not discuss the situation with my immediate district chief because I knew he would do nothing. I refused to accept this racial division of labor. I talked to Deputy Chief J.O. Barnett. Deputy Barnett worked on the opposite shift. He was approximately six feet tall, slim and military-like with a polite smile. I felt comfortable talking to him. I knew of him through a mutual friend, Mr. Joseph Garbarini Sr. I explained the situation to Deputy Chief Barnett. I emphasized that I was as competent as any officers detailed to that station. Chief Barnett said little but we maintained eye contact as we communicated. Later the district chief started rotating the details between the three lieutenants. I am certain officers at Station #6 were aware of my discussion with Deputy Chief Barnett. After my conversation with Chief Barnett, I had no problem being detailed at Station #6 and to other stations as well. In my philosophical father's wise words, I had moved a mountain by carrying one small stone at a time.

On one of my details to a fire squad, at approximately 9:00 p.m., we received an alarm. The call was to a high-rise apartment building on Poplar and Montgomery Streets. Several engines and truck companies arrived on the scene. My company entered the building and proceeded to the fifth floor in search of the fire. Through my air mask I could see smoke becoming thicker and blacker as we moved along, searching each room for the fire. Although my vision was tunnel-like, I saw streaks of light under the doors. It was sweltering in the building because of the intense heat, heavy firefighting gear we wore and equipment we carried. One of my men and I opened a door. A bright red glow radiated from a couch. The couch fire used all the oxygen in the room. The fire was now in the smoldering stage due to lack of oxygen. My company worked in pairs and our team's efforts paid off. We'd found the fire.

As strange as it may seem, when we fought fires, we automatically concentrated on teamwork. We had an unspoken way of protecting each other during a fire. I sensed that white firefighters felt the same. Color was no matter at this point. Perhaps it was because we knew the real enemy was the fire. Fire knows no prejudice. We concentrated on finding the fire. It was a matter of minutes before we had knocked the fire down. We made certain the job was completed and that the fire would not rekindle.

Once a black firefighter and a white firefighter were fishing on opposite sides of the Coro Lake. The black firefighter caught fish after fish, while the white firefighter didn't even get a bite. Finally the white firefighter yelled across, "Why is it you catch fish and I get none?"

The black firefighter thought a minute and then replied, "I guess on your side they are afraid to open their mouths." Some white firefighters had no limits in attempting to keep the races separated, especially when it came to social situation. Fear can cause people to act impulsively and do crazy things. An example of social fear occurred after a Christmas party at Private Nathan Kurts's home. The private invited everyone at the engine house to a Christmas party at his home. I was a lieutenant at Station #6. Private Kurts was a pleasant guy with a broad, friendly smile. He did not seem affected by peer pressure. He personally invited me and I accepted the invitation.

The day after the party, the covert conversation among the station personnel was that Driver Darden danced with Delores and I'd danced with Driver Darden's wife. The overt conversation included the idea of Private Kurts putting the invitation on the engine house's blackboard. The invitation had been on the board for a week and nobody complained. Suddenly the invitation on the board became an issue. All the captains at the engine house were for the party until they saw me there. Consequently, the real issue was that Kurts invited me to his home and I accepted his invitation.

According to the captains, both of us violated the traditional code of behavior regarding social contact between black and white firefighters. This was unacceptable in the eyes of the captains and they showed their disapproval.

Robert fights
fire with fellow
firefighters.
*Photograph
courtesy of Robert
Crawford.*

To show objection to Private Kurts's action, the captains retaliated; they refused to talk to Private Kurts and Private Darden. They also resorted to name-calling. They called Private Kurts and Driver Darden "nigger-lovers" because Kurts invited me to the party and because of the dancing. While no one called me a name to my face, I was certain they called me names behind my back. The captains tried to intimidate me by not talking to me and withholding information. Some of the privates were friendly and showed some dislike for the captains' attitudes. I felt they supported me, but they were afraid to speak out. They feared the possibility of retaliation from the captains. Nitpicking and name-calling continued at the station. Conversation and name-calling got so bad that we had to go before Larry Williams, Chief of the Department. Chief Williams ignored the petty complaints. He advised that no more private announcements were to be placed on the fire department's bulletin board. The board was to be used for department business only. Chief Larry Williams threw the case out. Subsequently, the captains lost favor with the men they supervised. They later put aside their differences.

Chapter 10
Whose Side Are You On, Anyway?

Memphis again had its share of activities during the latter part of 1971. Seventeen-year-old Elton Hayes died after a high-speed chase with the Memphis Police. In retaliation for the youth's death, blacks rioted in the city. It was like 1968 when Dr. Martin Luther King was assassinated. Fire bombings, rock throwing, shooting, arson fires, violence toward firefighters and damage to fire equipment rapidly ensued. After five days the city began to return to normal. However, trouble still brewed with the city because of disparity in pay between firefighters and policemen.

Meanwhile, black firefighters still struggled to gain fair treatment from the Memphis Fire Department. From 1955 to 1971, the number of black firefighters only increased to thirty-two. Wyeth Chandler was now mayor elect of the city. Chief Eddie Hamilton was chief of the fire department. Chief Hamilton was placed on vacation and given other City benefits. He officially retired from the fire department in 1974. In the meantime, Lieutenant Wallace, Driver Stotts and I met on numerous occasions with fire and city administrators regarding hiring and promoting minorities in the department. They repeatedly treated our concerns with little or no interest. We waited and we waited.

The wait was not by choice. Information was key; I continued observing management, reading books, and talking to those white firefighters who were willing to talk about the department and its operations. When the fire stations integrated in 1965, I felt black firefighters became more knowledgeable of the department's operations. We became aware of what was going on at other stations. I also became more aware of the difficulty some white firefighters experienced when they were not in "the network." A few of them had difficulty even though they may have been deserving of promotions.

It was still much more difficult for us to make promotion the way the department's promotional procedures were set up. One thing was certain;

they knew we were good firefighters. From the time we were hired, we were under a microscope. When you work under as much pressure as we had, you couldn't afford to make mistakes. We knew we were good firefighters when young white officers were sent to #8 for us to "take care of." We knew we were good firefighters when a difficult fire raged and the chief was glad to have us on the scene. We knew we were good firefighters when he'd tell us, "y'all go in there and put that fire out for me."

Yet some white firefighters spread misinformation that suggested black firefighters were afraid of fire and therefore could not fight fire. Some white firefighters continued attacking our firefighting skills. It was a psychological game to make us think we were worthless. The game failed because we were psychologically prepared by society in general. We had to prove that we were as good as and better than some white firefighters. Most white firefighters knew we were good, but many could not accept that fact.

The department was slow to hire more black firefighters. I felt we needed to organize to fight discrimination and unfair treatment we received on the job. Hiring, promoting and fair treatment of black firefighters on the job was an off-limits subject to administration. After fourteen years, we still had little more than false hopes and more promises from the administration. We took exams. Year after year we were passed over, even though we were eligible for promotion by grade score and by seniority. They continued telling us to "wait until the next time" or "we'll get you next time." "The next time" got farther and farther away. I'd thought about a black firefighter's organization. At times we can learn simple lessons from the smallest creatures of nature. For example, when spider webs unite they can tie up a lion. We needed strength. We needed to validate our concerns. We could do it only by combining our forces and resources.

We needed an organization for black firefighters. The idea of a black firefighters' organization stayed on my mind. I began organizing mentally, asking myself "what if?" Organizing, prioritizing, arranging and rearranging the important issues at the appropriate time came from past experience of working with different people and organizations. I learned to network as a member of Orange Mound Civic Club and as president of St Thomas Catholic Church Pastoral Council. Past mentors provided me with organizational skills. Thus, the ideas of a black firefighter organization became stronger in my mind in 1971. It was at this time that I began networking.

I talked to black firefighters in cities I'd visit. I gathered names, telephone numbers and information about hiring, promotions and treatment of black firefighters in cities I'd visited. Each black firefighter I talked to became a connection that eventually led me to other firefighters. I learned they were all experiencing the same problems with hiring, promotions and unfair treatment. Later, I would make the connections I had hoped to make.

Several years ago we'd met with black firefighters from small, predominately black towns in Missouri. They came to Memphis when we were at Station at #8. I already knew some of them. However, I continued making connections with black firefighter in other cities through black firefighters from Missouri. These firefighters continued coming to the Fire Department Instructors Conference until the FDIC was discontinued. The conference was discontinued because the Memphis Fire Department was not unionized.

Delores and I vacationed in New Orleans in 1971. I met Private Warren McDaniel when I visited a fire station in that city. Private Warren was of medium height and build. He appeared interested in his career as a firefighter. We discussed issues of discrimination on the job. We were having the same problems on the job with administration's hiring policies and unfair treatment of black firefighters. We maintained contact and provided each other information about firefighters in other cities.

The contact I needed came in 1972. Lieutenant David Floyd and I met through networking. Lieutenant Floyd was tall, brown-skinned, confident and diplomatic. He was president of the International Association of Black Professional Firefighters (IABPFF). We had a lengthy conversation over dinner at my home. The first IABPFF was formed at the convention in Hartford, Connecticut, October 1970. The national organization created a liaison among the minority firefighters across the nation. Its purpose was to gather and compile information regarding injustices and to take action to correct them. Other cities had begun organizing local chapters of the national organization. We learned that we shared common problems with other fire departments. Our problems, among others, included discrimination in promoting and hiring black firefighters. We had similar complaints about the written examination. We discussed how the performance examinations were too subjective. Lieutenant Floyd stressed the importance of local chapters working within the structure of the International Association of Black Professional Firefighters. He and I agreed to maintain contact and to continue networking.

Private Warren McDaniel came through Memphis in 1972 en route to St Louis. We met and attended the IABPFF Conference in St. Louis, Missouri. The Memphis Fire Department provided no financial support for me to attend this conference. However, many years later the City would provide financial support for black firefighters to attend black firefighter conferences in other cities.

Black firefighters from all over the United States attended the International Association of Professional Black Firefighters conference in St. Louis,

Robert with St. Thomas Pastoral Council in 1975. *Photograph courtesy of St. Augustine Catholic Church.*

Missouri. It was confirmed at this IABPFF Conference that black firefighters in almost all cities and states were experiencing problems regarding promotions, hiring, work details, disciplinary actions and racial acts of intimidation. The purpose and aims of the IABPFF included creating liaison with black firefighters across the nation for recruiting blacks for jobs and collecting and evaluating data pertaining to conditions in areas where minorities are employed.

I have read, "No organization is worth anything that does not preserve its history." I returned from that meeting in St. Louis determined to follow through with the idea of organizing Memphis black firefighters. I contacted Lieutenants Carl Stotts and Norvell Wallace. I briefed them on the IABPFF conference in St Louis. I discussed the idea of organizing Memphis black firefighters with them. They expressed interest in the plan. From this point on we were like spider webs uniting to tie a lion.

This same year Lieutenants Norvell Wallace, Carl Stotts and I held our first meeting at my home. At our first meeting we discussed promotion, hiring and the immediate challenges. How would we start this organization? Would we continue networking? What do we want to accomplish? What is our main objective? Other questions we pondered included whether we wanted to join the national organization. What name would we use for the local organization? We asked ourselves questions that we anticipated from future members. We were aware that we needed the national organization. However, we were also aware that we needed the voice of a local organization. Consequently, we decided to organize our local black firefighters. Before we adjourned that night we had answers to our questions. The main objective was the same as that of the national organization—equal and fair treatment from the fire department for minorities.

We continued the meeting with the election of officers. I was elected president, Lieutenant Carl Stotts was vice-president and Lieutenant Norvell Wallace was elected treasurer. The black firefighters had taken the first step in becoming an organization with a purpose. I felt comfortable as the first president of the newly organized black firefighter organization. Our next task was to recruit members for the organization.

The NAACP provided the support that we needed from the community. This organization meant a lot to black firefighters. We concluded it was important to maintain memberships in the local NAACP. We visited the association when the office was on Hernando Street. That organization was instrumental in the first black firefighters being hired by the City. They were supportive and helped us stay there. They helped us fight discrimination on the job. This was one of their main purposes. In 1968 we explained our situation to the organization. They confronted the fire department with complaints that no blacks had been hired or promoted above the rank of private. Of course, the department denied any bias in hiring and promoting black firefighters. We supported the NAACP with individual memberships and an organizational membership. They supported us by encouraging our efforts to seek fair treatment from the fire department.

Cities with black firefighters' organizations had chosen various names. There were the Vulcan Society of New York; the Fire Bird Society of New Haven, Connecticut; Van Guards of Cleveland, Ohio; Phoenix Society of Hartford, Connecticut; and the Valiants of Philadelphia, Pennsylvania. Some of the names that we considered included Blazers, Flames, Memphis Chapter of the Vulcan Society, Sparks, Bluff City Firefighters and Fire Blazers of Memphis. We would become the Pioneer Black Firefighters. Young black firefighters suggested "Pioneer" would also honor the first twelve black firefighters. The Maltese cross symbol with the number twelve in the center would represent the twelve original black fire fighters.

For the time being, we continued having monthly meetings at each other's homes until our membership grew. Although there were only thirty-two black firefighters in the department, each time we met we had a few more members. When we became too large to meet in homes, we moved our meetings to St. Thomas Catholic Church in South Memphis at 588 E. Trigg at Lauderdale Street. The late Archbishop James Lyke of Atlanta was pastor at that time. The Pioneers were becoming a reality. We moved from St. Thomas Catholic Church to Four Way Grill at 998 Mississippi Boulevard. They let us meet rent-free. The only stipulation was that we buy their food. We soon outgrew Four Way Grill. We met on Jackson Avenue at a storefront building. We met there until one of the members ran for political office. We went for a meeting one night, and our member had turned the meeting place into his political campaign headquarters. Instead of the Pioneers paying rent we made the "potential politician," Ulysses Jones, pay rent for the next two months. We left the politician's headquarters and held meetings at Newsum's Annex. For fundraising we held dances at the American Legion Hall. The Pioneers later leased and operated a nightclub at Vance and Front Street for meetings and public entertainment. The Pioneers held dances there to raise money to pay legal fees for its members. The Pioneers eventually discontinued the nightclub business.

After we organized the Pioneer Black Firefighters, other black firefighters joined the Pioneers. At one time we had 90 percent of the black firefighters in the organization. It was up to the members to convince others that what we were doing was to make it better for them as firefighters. We were aware that black City employees in general were experiencing the same problems with promotions. We decided to open the membership to anyone who worked for the City and wanted to join the Pioneers. Some joined and

Formerly St. Thomas Catholic Church. *Photograph courtesy of Robert Crawford.*

stayed for a short time. They let it be known they were not interested in the black firefighters' organization. Some black firefighters complained that we did not have the $25,000 life insurance that the union offered. We would not let that deter their joining the organization. Since Lieutenant Wallace sat on the credit union board, he pushed for dues automatically deducted from members' paychecks. At first the credit union refused the dues deductions. However, the credit union finally agreed. We drafted the letter authorizing the City of Memphis Credit Union to deduct from members of the Pioneer organization the sum of ten dollars each pay period to the Pioneer Black Firefighters Association.

Offering the life insurance policy helped increase our membership. As time passed, in addition to having majority black firefighters, the diverse group also included black female department employees. Fire department secretaries, clerks and typists were part of the group. Some maintenance mechanics also became members of the organization.

Ten years passed before the Memphis Fire Department hired the next three black firefighters in 1965. During this same year the fire department hired ninety white firefighters. Our goal was to change the pattern of promotion and hiring practice of the department. Promotions were even slower. The first black lieutenant, Norvell Wallace, was promoted in 1969.

Our lines of communication were tight. Lieutenant Stotts and I were on the same shift. Lieutenant Wallace was on the opposite shift. The three of us talked frequently. We kept each other informed of what happened on our shifts. For example, I remember the time we got word through the grapevine that there were plans for a credit union meeting. We heard the meeting would be held in the city council chambers. Past credit union meetings included no blacks. The Pioneer's plans included having black representation at the City of Memphis Credit Union's meeting. We decided that if we could have a good turnout for this meeting, we could elect a black to the credit union board.

To insure a good turnout, we contacted Pioneer Black Firefighters' members. We informed them of the meeting. We encouraged them to be there. We needed numbers to vote a black person to the City of Memphis Credit Union Board. We turned out, but when we got to the council chambers, word was out that we would attend the meeting. Consequently, it seemed as if the whole white police department was there. The situation was reminiscent of the 1960s sit-ins. There were so many people from the fire and police department that they cancelled the credit union board meeting. With help from the Pioneer efforts we later obtained black credit union employees and black members on the credit union board. Lieutenant Wallace and I were the first blacks to serve on this board. We'd then push for other black board members and finally we'd push for black employees in the credit union. Later, Mrs. Barbara Arnold and Mrs. Christine Campbell served on that board. I continued pushing the manager to hire more blacks in the credit union. The manager asked me to send him somebody. Darlene White was already working there. Lieutenant Wallace, Stotts and I continued making recommendations. Eventually blacks were hired at the City of Memphis Credit Union as clerks, typists and credit counselors.

In 1971, two different firefighter groups were organizing for two different reasons. The local Association of Firefighters organized as the result of the City refusing parity in pay with the police department. Black firefighters organized to fight discrimination and unequal treatment on the job. The local union invited everyone to join, including black firefighters. The union voted driver Carl Stotts vice-president. Driver Stotts, Lieutenant Wallace and I discussed our issues and concerns with union leaders and with administration. Since we were a part of that organization, we expected the organization to support our cause. The union, like administration, showed little concern for our issues. The union leaders later advised Driver Stotts that they could not help us with our problem. They could not side with us against the white firefighters .The union leader suggested that we start our own organization and they even promised to help us. This was just another empty promise from another source. The union showed no interest in our issues. They never helped us. But we would not stop building the house because the nail broke. We simply changed the nail.

Administration changed in 1973. Chief Hamilton was leaving office. Deputy Chief Robert W. Walker became director of the Division of Fire Services. At this time we had thirty-two black firefighters on the job. We always did our homework before meeting with either the fire director or the personnel director. Lieutenant Stotts, Wallace and I always got together to discuss what we wanted from the fire department. We were aware there were eight divisions in the department. Our goal was to have black representation in each division of the city's fire department. For example, we felt that a black firefighter needed to be placed in training. With a black firefighter in this position, we'd have access to information about what we needed to do to keep black firefighters on the job.

We immediately made plans to talk to Director Walker about hiring and promoting black firefighters. Director Walker—tall, slender, alert— appeared to think before he spoke. He seemed to be the kind of person who holds back when put in a new situation. He listened to us. He honored some of our requests. For example, he placed James Lewis in the training division as we requested. However, he would not honor our request for hiring and promoting more black firefighters, although he said he would. We were aware the director was getting his orders from city hall.

Gradually Director Walker started honoring more of our requests. It was encouraging that our organization apparently began to have effects on the Memphis Fire Department. I continued networking. I remained a member of the IAPBFF. Later, I became a member of the Black Fire Chief Officer Network.

That same year I'd returned from the International Association of Professional Black Firefighters Conference in St. Louis with the national symbol of the IABPFF that I purchased at the conference. The decal was a Maltese cross with a clinched black fist in the center. I placed the decal in the rear window of my car. I remember a chief stared at the decal, but said nothing. I felt that he told Director Walker about the decal. I am sure administration was aware that black firefighters were meeting. It appeared that after word got back to city hall that black firefighters were organizing, situations started to improve regarding hiring and promoting black firefighters.

The Pioneer Black Firefighters was established in 1973. By this time more cities had either started an organization or were in the process of doing so. In 1971 I'd contacted Vincent W. Julius, corresponding secretary of the Vulcan Society in New York. He sent me a copy of the Vulcan Society, Inc. FDNY constitution and bylaws. I received a letter from the regional president of IABPFF, Calvin Brown. He approved the use of the Vulcan Society's constitution as a guide for our own organization. Using the guidelines that I received, I made the necessary changes in their constitution to fit our organization. In 1975 we held an executive meeting to revise the constitution, bylaws and the charter. Pioneers attending that meeting included Private Sterling Phillips, Private Willie Flowers, Lieutenant Wallace and Driver Charles Wright. Another priority at this meeting was to seek new members. In 1977 we were chartered as the Pioneer Black Firefighters, Inc.

The Pioneer Black Firefighters were the first black non-union organization in the city; we were not a bargaining unit. We were never recognized as such. We wanted justice on the job. The union did not interfere with our starting the Pioneer Black Firefighters' organization. However, in their Memorandum of Understanding, they stipulated that they were the only ones who could represent firefighters. They made it clear that they represented white and black firefighters.

We discussed the union and agreed that the local union had black members in their organization and used black members' money to help fight against affirmative action. Of course they denied this, as the local union felt we were forming our own. Most black firefighters did pull away from the union. The union was not pleased with this. We explained to our members and to the union that our organization was a professional organization concerned with fairness in promoting, hiring and treatment of black firefighters. We knew we had no bargaining power. Our goal was to fight discrimination on the job. We wanted what the administration and the union turned a deaf ear to. The Pioneers organized to get fair and equitable hiring, promotions and fair treatment within the Memphis Fire Department. The organization put teeth in our demands. It would give us a voice in the workplace.

We were getting nothing done by administration and we were not going to get anything done by the union. Since our issues were ignored by both entities, we discussed filing a discrimination suit against the city. We decided to hire attorney Otis Higgs to represent us in a class-action discrimination lawsuit. We invited him to our meeting and explained what we wanted. He said, "You know I appreciate you folks for what you want but you want somebody up-to-date and knows more civil rights law than I do. There's a firm…Richard Fields is one of the attorneys. I appreciate you thinking about me, but give Richard a call."

We accepted Attorney Higgs's advice and contacted the law firm of Richard B. Fields, Esquire, Ratner, and Sugarmon & Lucas. We invited

STATE OF TENNESSEE
SECRETARY OF STATE
NASHVILLE, TENNESSEE 37219

GENTRY CROWELL
SECRETARY OF STATE
CAPITOL BLDG. 741-2816

JAMES P. BRADLEY
EXECUTIVE ASSISTANT
CAPITOL BLDG. 741-2816

March 3, 1977

ADMINISTRATIVE PROCEDURES		
580 CAPITOL HILL BLDG.		741-7008
CORPORATIONS		
C1-101 CENTRAL SERVICE BLDG		741-2286
ELECTIONS		
904 CAPITOL HILL BLDG		741-7956
PUBLICATIONS		
904 CAPITOL HILL BLDG.		741-7956
TRADEMARK SECTION		
CAPITOL BLDG		741-2817
UNIFORM COMMERCIAL CODE		
C1-101 CENTRAL SERVICE BLDG		741-3726

RE: THE PIONEERS- BLACK FIREFIGHTERS, INC.

Mr. R. J. Crawford
P. O. Box 26001
Memphis, Tennessee 38126

Dear Sir:

You will find enclosed the Charter for the above named, which was recorded in this office as of this date, together with a receipt for $15.00 and an overpayment check in the amount of $11.00.

Please cash this overpayment check immediately.

This office does not handle the registration of corporation work in the various counties since the enactment of Chapter 495, Public Acts of 1970.

Very truly yours

Gentry Crowell
Secretary of State

LG:Encls.

Charter for the Pioneer Black Firefighter Organization. This document shows the year the organization was officially chartered.

attorney Richard Fields to our regular meeting. This meeting with Attorney Fields included Robert Young, Norvell Wallace, Carl Stotts and John Cooper. We explained our plans to file a discrimination lawsuit against the fire department. He listened to our complaints. He agreed to represent us. Attorney Fields dealt with the City personnel director, CAO or chief administrator officer and the City's attorney. We kept Attorney Fields abreast of the problems we were experiencing with the department. For example, fire stations had syndicates whereby they pooled money to purchase food and everyone cooked and ate together. White firefighters at #8 refused to include us in the syndicate. Attorney Fields talked to administration about the problem. Administration told the white firefighters they would have to include black firefighters. Well, rather than conform, they broke up the syndicate. It was much later before we started the syndicate again. Meanwhile, Lieutenant Wallace, Private Stotts and I dealt with fire administration. We met frequently with Attorney Fields to exchange information and ideas pertaining to the class action discrimination lawsuit.

Chapter 11

The Pioneer Progress

O f the 343 Memphis firefighters hired in 1969, only 16 were black. Discriminatory practices in hiring and promoting black firefighters continued. There were very few, if any, black employees in the other divisions of the fire departments. It was a sad commentary for one of the top fire departments. It spoke volumes for the city of Memphis. Each year the Pioneers became more determined to end the fire department's inequity. The Pioneers would not give up the struggle. We simply couldn't.

In addition to fighting fires and discrimination on the job, the Pioneers were also becoming known in the community for charity work. The Pioneers initially started fundraising activities to fight discrimination in 1972. One of our first community interests was the Boys Club of Memphis. We presented annual scramble golf tournaments at T.O. Fuller State Park Golf Course. We also sponsored dances at Club Rosewood. This club was located in an old theater on Lauderdale Street in South Memphis. Funds were donated to support needy families in our community. Our fundraisers were successful.

As president of the Pioneers, I'd applied for a non-profit charter. Therefore, the organization was already tax-exempt. We solicited help from businesses in the community. People and the business community were quite responsive to many of our requests. I remember contacting a former employer, McDonald Brothers Co., Inc. The company gave me two TVs and two radios to use as prizes in our tournament. All members of the organization worked diligently to make the Pioneers successful in assisting the needy in our community. For example, the Pioneers made financial donations to families who suffered loss due to fires. While we continued helping the needy in the community, the Pioneers were forced to focus on another venture that required financial attention.

A few closed doors began opening for us in 1972. However, we had long ago ventured on a journey that would take different twists and turns during many years of fighting for fairness. Our attorney, Richard Fields, more than matched our efforts in working meticulously to protect our rights. He ignored absolutely nothing regarding our civil rights. Our concerns were his concerns. For example, the Pioneers were concerned about William Carter, John Cooper, George Dumas and Floyd Newsum returning to the department. Chief Eddie Hamilton fired William Carter for insubordination in 1964 during the St. Augustine incident. He promised that Carter would not return to the department as long as he was in charge. John Cooper was suspended for cursing in the fire station. He was dismissed after he called a witness and admonished the witness to tell the truth. Floyd Newsum left the department after he was refused a leave due to personal problems following the assassination of Dr. Martin Luther King. I think that George Dumas became disenchanted with the department in general and left. Murray Pegues relocated to Los Angeles, California.

White firefighters also had their reasons for leaving the department, but they did not have to hire an attorney to return to work. The City appeared more than willing to rehire the white firefighters. The Pioneers' attorney Richard Fields argued the case for black firefighters who wanted to return to the department. "If white firefighters return to work, then so should the black firefighter." With efforts from the Pioneer Black Firefighters and Attorney Fields, black firefighters who had left for various reasons returned to the department.

Meanwhile, the Pioneers continued their efforts in fighting discriminatory promotions and hiring practices in the department. We also continued our fundraising activities to cover attorney fees. We'd hired attorney Richard Fields to represent us in court. We laid out our problems. Attorney Fields was ready to take our case to court. We did not have all the money to pay the retainer fee. We needed additional funds to cover attorney fees. While Fields was working the lawsuit, Lieutenant Wallace, the treasurer, was working the budget. The Pioneers needed a $5,000 retainer fee. This was a lot of money at that time. The Pioneers continued working relentlessly trying to raise the money.

In 1973 a Pioneer was injured during a fire at the Memphis Moldings Company on Walnut Street in South Memphis. Lieutenant Stotts was commander of Station #8 pumper truck. They arrived on the scene and knocked the small fire down in a matter of minutes. Lieutenant Carl Stotts, making a second check to make certain the fire was out, opened the door to the blower room. The dust explosion lifted him, spun him around and knocked him to the floor. The force of the flames shot under the face shield and snapped the helmet off his head. He was seriously injured from the fiery explosion with burns to most of his body. Lieutenant Stotts was transported to Baptist Hospital. After forty-two days in the hospital, he recovered and returned to duty. His return would prove to be a good thing.

The Pioneers encouraged all black firefighters to keep a grade on file so that we would be prepared for promotions when and if the opportunity for promotion presented itself. No one could say we had not taken the test, that we were not prepared or that we were not qualified. We stood ready to support each other when necessary.

I remained in firefighting until I was promoted to fire investigator. Promotions and hiring had been frozen. This was the first promotion made in four or five years. Earlier, the NAACP complained about the fire department's discrimination in hiring and promoting blacks in the department. In 1976, Director Robert W. Walker promoted three black firefighters and three white firefighters. Lieutenant Wallace and I were promoted to fire investigator in the fire prevention bureau. Lieutenant Stotts was promoted to captain in the fire fighting division. Twenty-one years passed before the three of us were promoted to this rank.

Fire investigator duties included investigating to determine the causes of fires and where and how fires start. Qualifications included knowledge of fire department rules, regulation and orders, fire laws and ordinances, as well as general knowledge of municipal fire prevention and building codes. As fire investigator I learned to observe fire, to analyze cause, and to discover its origin. If the fire appeared accidental, the report was simply filed. I submitted reports of my findings to the arson squad if the fire appeared to be arson. I remember an early-morning fire that claimed the life of a mother and two children. Another firefighter and I investigated the fire after shift change. We took pictures of the fire, the victims who perished in the fire and the location where they were found in the house. We interviewed occupants of the house. We would have to determine if the fire was arson or accidental. We also reported on the appearance, contents, and building loss. We would then make our report.

Upon discussing the fire and the people we interviewed at the fire, the other investigator and I discovered discrepancies in the information the occupants gave us. We returned to the scene the following workday. A relative that we interviewed stated, "When I saw you coming back, I sensed that you did not believe me yesterday." She then told us the truth. She said, "A mentally retarded relative started the fire. I was afraid she would be arrested for setting the fire." We saw the relative who caused the fire. There was little doubt that the she was indeed retarded. We treated the fire as accidental rather than arson.

An antique trolley serves downtown attractions and a pedestrian shopping mall. It runs along the river and beyond. *Photograph courtesy of Robert Crawford.*

In 1974, a twenty-nine-year-old Memphian, Harold Ford, upset incumbent Dan Kuykendall to become the first black U.S. Representative from Tennessee. Meanwhile, the Pioneers continued their legal battles. Between 1950 and 1976 the department hired 94 black firefighters compared to 1,683 white firefighters. In the same year the fire department promoted forty firefighters to drivers; only one of the forty firefighters promoted to driver was black. The department promoted twenty-one lieutenants; none were black. Of twenty-eight firefighters hired in the same year, only three were black. This had been the pattern since we were hired in 1955. Patterns such as these were among other reasons the Pioneers organized and continued to fight unfair practices in the fire department.

We realized administration worked from the premise that every black firefighter hired and or promoted meant fewer jobs for white firefighters. The Pioneers worked from the premise that black firefighters in supervisory positions meant more opportunities for blacks in the department.

Nineteen seventy-four saw the beginning of changes in the fire departments in Memphis and other cities. In 1974 approximately 6 percent of 1,600 in the Memphis Fire Department were black. The plan was to increase the percentage of black firefighters. In the 1974 consent decree, the City promised to fill at least 50 percent of all City vacancies with qualified black applicants. This included the fire department. The City and the local union signed an agreement to speed up efforts to recruit black firefighters. This effort was funded by a labor department grant. The program was designed to prepare minorities to pass written, oral and physical agility tests that were parts of the firefighter civil service test. This announcement was actually made three years before the program started. In 1975 the personnel director said if training classes for the upcoming year were excluded from the City's budget, the money would be lost.

Meanwhile, the Pioneers continued struggling to raise attorney fees. However, we agreed to assess everybody one hundred dollars. Lieutenant Wallace kept a list of those who did and those who didn't pay the assessment. Problems arose when we needed more cooperation from the members. Some members paid their assessment and later requested a refund. On top of everything else, new members who had recently joined the organization dropped out and requested refunds. Meanwhile, Attorney Fields continued with the lawsuit. Those left in the organization continued working to raise attorney fees.

Since the Pioneers were unable to pay the attorney, Lieutenant Carl Stotts paid the initial $5,000 retainer. February 16, 1977, Lieutenant Carl Stotts filed a class-action suit against the City, the fire department and the director of the fire department. The suit alleged that the hiring and promotion policies of the fire department were racially discriminatory. From that point on, the Pioneers, through fundraising activities, paid attorney fees for black employees to fight cases of discrimination on the job. Later, the Pioneers would attempt to join the lawsuit.

Lieutenant Carl Stotts made his mighty move for all black firefighters now and in the future. Following Lieutenant Stotts's suit, we learned that white firefighters had a "cheat sheet" to fire department promotion examinations. The Pioneers had long suspected this practice was going on. We knew that certain white firefighters did not have high school diplomas. Wallace said in jest, "If you talk to some of them long enough, you'll know they had no high school education and if you look close enough you may notice some signing their names with an "X." Administration stipulated a high school education or GED after we came on the department. Prior to 1955, white firefighters did not have to hide the fact that they did not have a high school education, because no one challenged them.

Nineteen seventy-seven was a significant year for black firefighters. The Pioneers were ahead of other black firefighter organizations nationally in fighting discrimination because Stotts filed a discrimination suit against the City. This lawsuit opened doors for other black firefighters' organizations. Memphis was one of several cities chosen to receive a Labor Recruitment Grant. The program, through the International Association of Firefighters,

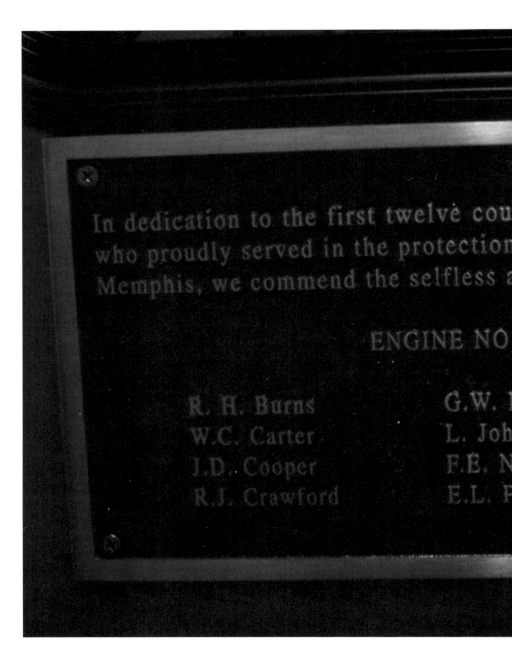

Engine #8 displays names of the first twelve black firefighters. *Photograph courtesy of Robert Crawford.*

was developed to help recruit applicants for firefighting jobs. This pre-application training program for potential firefighters placed emphasis on educating minorities.

In addition to educating minorities applying for jobs with the fire department, the program also provided tutors for the recruits. Tutors were expected to present the material in an understanding manner. They were expected to know the material to be used in hiring and to explain the scope of the program. White firefighters had relatives and friends to familiarize them with procedures of the fire department; black firefighters had none. We were motivated to help each other.

Classes were held at Newsum's Rehabilitation Annex to help future black firefighters pass the test. Among the 385 applicants there were 95 blacks. The Pioneers initiated the recruiting. We tutored the recruits so that they would meet the department's requirements. We helped the recruits with physical conditioning to meet standards of strength, endurance and agility. We held orientation classes to explain to the applicant the material used in classes, the extent of the program, and applicant expectations and department expectations. During tutoring, the applicants were introduced to materials they might encounter on the civil service examination. These classes lasted approximately nine weeks, four days a week for three hours a class.

I received a letter from the International Association of Firefighters Labor Recruitment Program dated September 15, 1977. The project administrator acknowledged the success of the recruitment program. Although the letter was addressed to me, it was in reality to all the Pioneers. No one person can accept credit for the success of the recruiting efforts of getting more blacks on the fire department. This has always been the Pioneer's goal.

December 1977 saw seventy-three recruits graduate from the fire-training academy. Of the seventy-eight recruits in the class, forty-seven were black and thirty-one were white firefighters. Five men did not complete the program. This was the first time in its history that the Memphis Fire Department recruited and hired more blacks than whites in the department. This was the first class to graduate in three years.

While the Pioneers continued efforts to get more blacks hired and more promotions for those already with the fire department, Lieutenant Stotts, Lieutenant Wallace and I had agreed on our goal. Our aim was to continue efforts in getting blacks employed in all areas of the fire department. We had much catching up to do. Many Pioneers have no idea of the things we did behind the scenes to support hiring and promoting blacks in the department. We worked behind the scenes in many situations to help pull other black firefighters up. Although we pushed for blacks, we looked for

Robert is presented with an award for dedication and service to the Pioneer Black Firefighters. *Photograph courtesy of James Mitchell.*

those who showed initiative. We didn't want to push anybody who obviously didn't want to work.

Since Lieutenant Wallace and I were on the credit union board, we continued to use our influence pushing for other blacks to be hired in the credit union. We continued to talk to the credit union manager about hiring blacks in that department.

I moonlighted as a licensed journeyman plumber when I was not working at the fire department. I purchased most plumbing supplies at Central Hardware. Debra Gist worked as the head cashier at that store. She appeared competent in her job. She was conscientious, courteous and intelligent. I was impressed with her knowledge of the job and her apparent polite attitude toward customers. Later, I talked to the manager about hiring a black person in the City of Memphis Credit Union. He simply said, "Send me somebody." I recommended Debra. She was hired and became one of the first black credit counselors at the credit union. Meanwhile, Lieutenants

Wallace and Stotts continued working on efforts in getting blacks employed in other areas of the department.

I felt it was time for me to give up the presidency of the Pioneers. It was time for new leadership. The young Pioneers needed the experience of leading the group. I discussed my feelings with Stotts and Wallace. Consequently, I did not run for office. At the November 10, 1977 meeting Lieutenant Jessie Jones was elected and became the second president of Pioneer Black Firefighters, Inc. Ulysses Jones was nominated vice president and Wallace remained treasurer. I became one of the trustees. Other trustees included Carl Stotts, Robert Moore, Terrold Bilbrew, Jessie Dorsey, John Alsobrook, Floyd Newsum, Robert Freeman and Benny Shields.

Chapter 12

You Are Not a Tree

The concept of racial quotas came under fire in 1978 when the Supreme Court ruled in the case of *Bakke v. the University of California*. Allen Bakke, a white student, claimed his civil rights were violated when he was denied admission because of racial quotas designed to increase the number of minority students in the medical school. In the same year, Jill E. Brown of Baltimore, Maryland, became the first black woman pilot for a major U.S. airline.

In Memphis, the weather ruled in 1978. It was a cold January. It took me back to my childhood and those cold hog-killing days. Memphians skidded, slipped and shivered through snow and ice the entire month of January. The fire department, as usual with cold weather, received more fire calls because people, trying to keep warm, become less careful. Single-digit temperatures with ice and snow hampered efforts of the fire department in terms of response time.

The National Democratic Party held its conference in Memphis. U.S. Representative Harold Ford, delegates to the convention, and reporters took an early-morning tour of the city. The tour ended at a Dunkin Donuts shop where a grand donut-throwing fight ensued.

Neither the grand donut fight nor the weather equaled battles the fire department experienced with black firefighters and the firefighters' union. Between 1977 and 1979, two burning issues proved problematic to the department. One of the issues was Carl Stotts's filing a lawsuit and the other issue was the firefighters' union strike. Attorney Richard Fields continued representing black firefighters' persistent problems with the department's hiring and promotion policy. Black firefighters discussed the idea of the Pioneer Black Firefighters filing a class-action suit against the

City. However, Fields advised the Pioneers to let Carl Stotts and Fred Jones file the suit against the City and the Pioneers join to make a class-action suit. Consequently, Judge McRae denied the Pioneer's motion to join the suit. Nonetheless, Stotts and Jones won their cases. On February 16, 1977, Stotts filed a suit alleging that he and other blacks were denied promotions because of their race. Similarly, in an individual action, Fred Jones alleged that the department had not promoted him to fire inspector because of his race. The two cases were eventually consolidated.

The other problem issue for the Memphis Fire Department was the firefighter's strike in 1978. The walkout occurred after the union and the City failed to reach agreement over wage disputes. The strike lasted three days. Striking firefighters refused to answer alarms. Many firefighting companies were out of service due to the strike. Management-level personnel, with the help of non-strikers and national guardsmen, manned the few companies remaining in service.

I was a union member in 1978. I'd joined the union strictly because of the minority recruitment program. Although I'd left my name on the union roll, I did not go on strike. One reason among many, I was fire investigator and captain and part of management. Union members tried to intimidate me. For example, a watermelon, left by strikers, was placed in the driveway of the station where I was assigned. It had "Robert Crawford" scrawled on it with "Soul Brother" scribbled under the name. After the strike, the union threatened to send me before a board that would determine my fate because I'd refused to strike. There was no way I was going to let a bunch of privates "determine" my fate. Absolutely no way. I was finished with the union, permanently. Even so, I never tried to influence black firefighters' decisions about the union, but I cautioned them to be mindful and stay out of trouble.

I have never been uncomfortable around or intimidated by white people. I was not taught to be inferior or vengeful. I will admit I am an intense taskmaster with high expectations, a stickler for rules. An old saying from my father was, "If you don't like where you are, change it! You're not a tree." I recall, when I was a lieutenant, asking for a transfer because I wanted my own command. The chief transferred me to a command position at a station in Frayser. The men that I supervised there were white except one black firefighter, Private Lloyd Moore. This was all right by me. My confidence was not brittle. Nonetheless, challenges from individual white firefighters came in many forms. I recall the subject of school busing coming up at one of the stations where I was assigned.

A fire chief who always avoided eye contact when he spoke to me said, "We ought to blow up all the buses."

I said, "Now, you're not the only one paying taxes on those buses. I pay taxes too. I own a wheel or something on those buses. I have some say about what should be done to those buses." The things that stand out in my mind about this chief are that he usually had a cigar in his mouth and a coffee cup in his hand. He was very vocal and sarcastic about busing. I remember always firing right back at him. He made it clear to me that he didn't care for blacks in general and me in particular. Some years later, he would confirm my suspicions. Nonetheless, I didn't spend my time looking for approval from him. I only wanted to be treated fairly and with respect.

Not all white firefighters displayed negative attitudes toward black firefighters. Some white firefighters were encouraging. For example, Chief Hershel Dove was a source of support. He appeared impressed with my work and leadership. Although he never verbalized it, I felt by his action that he realized my potential. In addition to being supportive, Chief Dove also encouraged me in other ways. For instance, when he was on the promotion interview boards he'd asked me to serve on every review board with him. My responsibility on the interview boards included questioning and grading applicants. Chief Dove appeared to respect my talents as a member of the fire department. I respected him for his knowledge and individuality.

The NAACP held a conference with the Pioneers in its office at 588 Vance Avenue, the organization's Memphis office. Maxine Smith, energetic, dynamic and vocal on civil rights issues, was executive director of the organization. Stotts and other members of the Pioneers attended that meeting. Attorney Fields found that "a disproportionate number of black candidates failed the test." He said it also became apparent that some white firemen had "increased their scores significantly between 1976 and 1978, some by as much as forty-nine points." Maxine Smith expressed her belief that the City knew about the cheating. The final decision was that the administration decided to redesign the test so that the cheat sheet would be ineffective. Even so, there were anonymous charges that black firefighters were given copies of the lieutenants' promotional exam. Pioneer president Robert Young's personal investigation revealed nothing to substantiate the allegations. He felt an investigation was the fire department's responsibility unless the department attempted to falsely accuse an individual. The Pioneers took no action on the allegation.

By 1979 the fire department was 10 percent black. The entire city was operating under the consent decree of 1974. In this decree, the City promised to attempt filling at least 50 percent of all City vacancies with qualified black applicants. This included the fire department. Plans for promotions in 1978 had been cancelled because copies of the written examination for promotions were stolen. To make matters worse for the fire department, the

U.S. Justice Department was in the city to approve the testing procedure and to review results of the tests already given. The justice department participated in the City's promotion process because of the consent decree of 1974 resulting from the discrimination suit filed against the City by the U.S. Attorney General.

The last promotions were in 1976. Consequently, 1979 held the largest group promotion in many years. This year marked another turning point in my firefighting career. The department promoted 197 firemen. Of the thirty-three district chiefs in firefighting, I was the first and only black firefighter Director Walker promoted to district chief. As district chief, I was in charge of five stations that consisted of approximately forty men. I was responsible for day-to-day operation of these companies. My duties also included, but were not limited to, firefighting, coordinating and supervising, planning and coordinating work schedules for fire stations and concurring and assisting in disciplinary counseling. Additional responsibilities entailed commanding initial emergency incidents. This included hazardous-material incidents. Other tasks included rating forms, doing evaluations and assisting in promotional examinations.

Being first black district chief certainly carried its own tribulations. It was no revelation that most white firefighters did not want black firefighters as supervisors. Some of them didn't hesitate to make you aware of their feelings. A few black firefighters didn't want a black supervisor, either. Discipline and experience in the military prepared me to cope with attitudes on the fire department. I returned to Station #6 in the North Memphis and Frayser area as district chief. This is the area that I'd served as lieutenant. My first fire as a Battalion Chief assured me the "good old boy" network still worked.

I remember challenging the "good old boy" syndrome and a lieutenant that I supervised. This lieutenant was not taking command of his company. I observed him to be one of those laid-back lieutenants. For example, he made a house fire. He left the fire and it rekindled. This is definitely a no-no. His job as lieutenant was to be certain the fire was out before he left the scene. Any time a fire rekindles it causes more damage. I talked to him about this incident and his lack of leadership with the men. He knew what he'd done was bad firefighting. He flashed his dirty little grin. He was obviously testing me. He acted like he didn't give a shit about what I had to say. He gave me that "to hell with you" stare. Apparently, he was not taking me seriously. He seemed to be telling me that I couldn't tell him what to do and that he would run things his way. He apparently saw the serious expression on my face. He tried to explain his way out of the situation. He knew rekindles were definitely unacceptable. The lieutenant tried to talk me out

Deputy Chief Robert Crawford. *Photograph courtesy of the Memphis Fire Department.*

of writing him up. Following the counseling session, I returned to my office and prepared a written reprimand. Consequently, he complained to Chief Dove. Later, Chief Dove rescinded my reprimand. If a centipede loses a leg, it does not prevent him from walking. The chief's rescinding this decision would not in the least deter my future actions.

I planned to continue making what I felt was the best decision in any situation. Despite Chief Dove's action, the lieutenant straightened up. In addition, word got out that "Chief Crawford don't take no mess." He wrote a letter on Lieutenant Rice, one of the "good old boys." Needless to say, this lieutenant and others got the word that I would not hesitate to "write a letter" when necessary. Nevertheless, I would encounter this "to hell with you" look later under different circumstances.

Other areas of contention included evaluating the men. Grades were based on seniority, written and performance examination. Supervision had nothing do with the written examination and performance. What I'd graded the "to hell with you" lieutenant on was performance. This lieutenant requested a promotion to captain. The "good old boy" expected a higher performance rating, no matter how poor his performance. He was disappointed with his performance grade. He complained that a lieutenant in another battalion received a higher rating.

"Well," I said, "I don't grade that way." I went over the rating form with him, pointing out his weakness and his strengths. I explained to him that I did not rate according to what others did. I explained that my rating was done on *his* performance. This lieutenant got it together and made two ranks under me. He respected me, he said because of my fairness.

Not all situations with resistant white firefighters ended amicably. Such expectations never entered my mind. Hardcore white firefighters that I found absolutely intolerable to work with, I transferred to other companies. Chief Dove usually supported me in these efforts by assisting in finding new assignments for these firefighters. It seemed that some white firefighters resisted black firefighters in every manner they could think of. Accordingly, these firefighters found it extremely difficult to have a black supervisor. They'd fought to keep us from making promotions and they fought the idea after it became a fact. Some openly fought the idea.

We had few in administrations setting good examples for men they supervised. I believe that leaders set the standards. Behavior filters down to the lowest man on the totem pole. Black firefighters were more likely than white firefighters to be chastised for the least infraction. Eyebrows rose and the grapevine buzzed for days when a white firefighter was chastised for an infraction against a black firefighter. The rare occasion happened early in my career as district chief.

The Tom Lee Memorial to African American Tom Lee, who saved thirty-two people from drowning, even though he could not swim. *Photograph courtesy of Robert Crawford.*

This rare occasion occurred when a district chief asked me to check on one of his men on sick leave. This firefighter had been on sick call for some time. I visited the firefighter's home. He was not there. If you are on sick leave you are supposed to be at home. I reported this to the firefighter's district chief. The district chief wrote a suspension letter on the firefighter. The firefighter appeared before the deputy chief for disciplinary hearing. I was there as a witness that he was not at home. However, I was unaware of the call made to the deputy chief until it was mentioned at the hearing. The firefighter called the deputy chief and asked, "Why did you send that nigger chief over here?" The deputy chief said, "I will not tolerate you calling the chief a 'nigger chief.' He is Chief Crawford." The firefighter was suspended for not being at home and for disrespecting an officer.

As with other professions, the fire department was not without its gossip. I was surprised to learn through gossip that a black firefighter called me "uppity," among other things. Thus, some white firefighters were not alone in displaying negative attitudes toward my position as supervisor. Now the department still did not have that many black firefighters and the negativity of a few would not deter the Pioneers. We continued our mission of getting more blacks hired. We continued giving input into black firefighters seeking promotions.

I had several conversations with another Pioneer who was extremely vocal about everything and appeared equally rebellious about everything. Our views differed on many matters regarding black firefighters' attitudes about our situation and the approach to take in solving problems we encountered. Nonetheless, I felt there was a place for all of us with the department. I also felt a little rebellion now and then was a good thing. But some of these guys stayed in perpetual motion. Every black and white firefighter had his way of dealing with the system. I tried to ignore those who criticized mine. Booker T. Washington said, "These are two ways of exerting one's strength: one is pushing down, the other is pulling up." I strived to do the latter. I never intended to retire from the fire department a bitter old man wallowing in regrets and self-pity.

Chapter 13

Talk Little, Learn Much

The court ordered copies of the consent decree of 1974 posted at all fire stations. District chiefs were ordered to post copies at their stations. I posted copies at my stations. No one initially filed an objection to the decree. However, a group of white firefighters filed a countersuit after the fifteen-day deadline. They felt the decree made them victims of reverse discrimination. U.S. District Judge Robert M. McRae refused the white firefighters' request to change the consent decree concerning promotions for black firefighters. He said, "I don't see how in the world any fair promotions compensate for any past failures to promote minorities…that won't help the majority from losing some spots." He added, "That's why they had the system in the first place." White firefighters claimed they knew nothing about the suit. The court ruled, "They waited too long to come forward."

At the beginning of 1980 in his letter to the Pioneers, attorney Richard Fields informed us that the lawsuits filed by Carl Stotts and Fred Jones alleging racial discrimination in hiring and promotion in the Memphis Fire Department were settled. He also informed us of our rights under the settlement and the opportunity for us to file objections to the settlement. The notice stipulated that the lawsuits were not only on behalf of the original plaintiffs but also on behalf of present and future black employees. This included all black persons who were denied employment since March 24, 1974, and the City consent decree date. In addition to changes in hiring and promotion practices, the settlement provided back-pay awards based on seniority and attorney fees. In his letter attorney Fields stated, "This settlement would not have been possible if the Pioneers had not stuck together as an organization. Only with unity of purpose were we able to show the City of Memphis our determination to succeed."

In the letter Attorney Fields gave special recognition to Carl Stotts and Fred Jones, who filed the initial charges. He also gave special recognition to the executive board of the Pioneers, and the two presidents, Jesse Jones and Robert Young who provided leadership during the suit.

Our fight for hiring and promotions did not end at this point. The Pioneers continued pressing the City to hire and promote not only minority firefighters but also other minority personnel within the fire department. We continued efforts to get blacks in all bureaus of the fire department.

Other issues we fought included the City budget, layoffs and rank reduction. In 1981, due to budgetary problems the City planned to lay off personnel or reduce their rank. Thirty-one employees were asked to voluntarily accept layoff or reduction in rank. These employees included seventeen black lieutenants and seven black drivers. Now there were approximately 225 white lieutenants and 300 white drivers who would not have been affected by the layoff or rank reduction. Since layoffs were supposed to be by seniority basis and because blacks were the last hired, it was obvious we were slated to be targets in the plan. Through this layoff and rank reduction we saw our gains invalidated.

As the department faced problems with layoffs and seniority issues, Director Robert Walker retired. James Smith replaced him. Director Walker was soft spoken and direct. Director Smith was vociferous, steadfast.

Robert Young, paramedic supervisor, was president of the Pioneers at this time. Even though only 11 percent of the department was black, we were pretty well situated in terms of black firefighters' participation with the Pioneers. However, we continued efforts to increase membership. We remained a persistent organization. Dedicated and committed, the Pioneers realized the layoff and rank reduction of black firefighters would destroy affirmative action that the court granted in the consent decree. We were pleased with Judge McRae's decision following the court hearing. In the court of appeals in Cincinnati 1982, Judge McRae's decision was upheld. He'd ruled that the City of Memphis couldn't meet the budget by laying off or demoting minorities.

The Pioneer Black Firefighters were taking care of business in litigations in 1981. Black firefighter union members were in a conflicting role. Here they were covered by the union contract with the City. At the same time they were at odds with the very union they supported. The City was obligated by the union's contract that said layoffs should be by seniority. The court ruled the seniority system discriminatory. The Pioneers were also kept busy in court by the department's residency requirements. The Pioneers said the City administration's policy of hiring only Shelby County residents was not enforced from March 1977 and August 1980. The Pioneers challenged the

City's hiring policy. The organization felt the union had done a poor job in fostering race relations with respect to affirmative action. For example, the union fought the first group of black promotions and later fought for a seniority layoff plan that would have had a negative impact on black firefighters. In addition, the union trial board fined Robert Young and other black firefighters six hundred dollars and six months suspension because the Pioneers filed a lawsuit in Chancery Court asking that City employees live in Shelby County. Consequently, black firefighters left the union and joined the Pioneers.

In the midst of the seemingly chaotic situations occurring in the department, the Pioneers maintained it was not a union. The Pioneers continued protecting rights of black firefighters. Strange as it may seem, firefighters took on a different demeanor when fighting fires. Black and white firefighters continued functioning as a team. I never heard of a situation where a black or a white firefighter felt alone while fighting a fire. Everything appeared secondary to firefighting and protecting lives, others and ours. Therefore, teamwork remained a priority even though we disagreed intensely on present issues. In every sense of the word, I truly feel that firefighters, black and white functioned as a team when it came to fighting fires. I remember answering a fire call at an apartment on Bickford Street in North Memphis. Two pumper companies and a truck company were on the scene when I arrived. The upstairs apartment was well involved. The fire had vented itself out the window of the north bedroom and living room. The entire contents of both rooms were consumed by fire. The other rooms of the apartment were charred and filled with smoke.

People excitedly shouted to me, "A child in there. He still in there, y'all!" One pumper laid the booster for search-and-rescue purposes. The second pumper had laid a two-and-a-half-foot line. The heat was intense and the smoke was heavy and dense. I radioed other fire companies regarding necessary action to take when they arrived on the scene. They immediately began searching for the missing child. We worked as a team with one goal. We had to find two-year-old James Mathews. You could hear it in their voices and see it on their faces, the sheer exhilaration when we rescued that little boy.

I maintained contacts with the International Association of Black Professional Firefighters. I attended the seventh biennial convention in August 1982 in Lexington, Kentucky. Using my days off and my own money to attend this conference, I addressed a committee regarding our accomplishments in Memphis. By 1982 the Pioneer Black Firefighters in Memphis had made more progress than any other black firefighter organization nationally because of Stotts's lawsuit. Consequently, Stotts's

Memphis Fire Department in action. *Photograph courtesy of the Memphis Fire Department.*

lawsuit set the precedent nationally for future discrimination lawsuits in other fire departments.

In 1982 a winter freeze with record low temperatures hit Memphis. We were in for a cold, hard winter. In Chicago, John Cardinal Cody, Roman Catholic Archbishop since 1965, died. He was an outspoken advocate of integration. That same year sports fans mourned the passing of Leroy "Satchel" Page, one of the great pitchers of all times in the Negro league. In Memphis, Mayor Wyeth Chandler resigned as mayor to accept an appointment to become a circuit court judge. City Council Chairman J.O. Patterson Jr. was sworn in as acting mayor, becoming the first black in Memphis history to hold the mayor's job. Patterson lasted only twenty days when an election for the position of interim mayor was held. The fire department was beginning to hire and promote more black firefighters.

The Pioneers remained steadfast and committed to the fight for hiring and promoting black firefighters. We kept administration on its toes because they apparently wanted to disregard the consent decree. Court appearances were uncommon before blacks were promoted or hired in the department.

Administration had to be reminded of the consent decree before I was promoted to deputy chief in 1982. I had experience, I had seniority and I had scored 85 on the test, the highest of the contenders. I applied for the deputy chief's position. However, when it was time for promotion, Director J.R. Smith and personnel director Joe Sabatini called me to the City personnel office. Director Smith appeared law-abiding. However, if he disagreed with the law, he didn't hesitate to strongly voice his objection. Joe Sabatini generated options, made choices and followed through. I felt being summoned to meet with these administrators probably meant a discussion regarding promotion to deputy fire chief.

Director Smith told me that I would be promoted deputy chief "next time." They wanted to promote a white district chief to that position. Apparently, they expected me to go along with the idea. I felt my agreement would have served as a possible excuse for them to say, "We offered him the position but he refused it." I was more than well aware of the orders in the consent degree.

The 1980 consent decree stated that the fire department would promote 20 percent minority in each rank beginning July 1 of each year. According to the decree, a black would get the promotion. I listened intently as the directors gave various reasons why the promotion should go to a white district chief. I don't know how long the meeting lasted but I think it was short. I listened to the "sweet talk" as I engaged my mind reflecting past offers prior to the consent decree to "wait until the next time." I also reflected Attorney Fields's message from personnel prior to Stotts's lawsuit. The message from administration was the department would promote Stotts, Wallace and me if we didn't push promotions of other black firefighters. We didn't accept the deal then and I refused to accept it now. I knew they were not promoting according to Judge McRae's order under the consent decree. It was time to seek legal counsel again.

Two things were abundantly clear about administration and the politics that go with the fire department; the mayor of the city and the director of the fire department were in agreement when it came to affirmative action. They felt it was wrong but they realized they could do nothing but obey the law. My answer was short and to the point. I didn't agree to "wait until the next time" to be promoted as suggested. I was a twenty-seven-year veteran of the Memphis Fire Department, I had a passing score of 85 and I had the experience and confidence. I was qualified for the job.

Following my meeting with administration, I decided to help them obey the law. I contacted Richard Fields. We discussed the meeting I'd had with the directors regarding the promotion.

Attorney Fields consulted Judge McRae. Our attorney and the Pioneers were in action again torch-bearing for black firefighters. The matter regarding black promotions was back in court. Many Pioneers attended the hearing regarding the potential promotions. There would be thirty-three promotions for the fiscal year 1982. We wanted at least seventeen promotions for blacks. We asked for a ruling regarding blacks being overlooked in the proposed promotion of lieutenant and above. All of the consent decree's requirements were met, except for deputy chief and district chief positions. Director J.R. Smith testified that he made the decisions on the rank following the applicants' tests scores and interviews. He wanted to promote District Chief Elvis Lemon to deputy chief because the director thought that Elvis Lemon "could do the most things best." Judge McRae called the Director's method "too subjective to pass the test of validity."

Attorney Fields's argument that compliance with the consent decree required at least 20 percent of promotions in each category to qualified blacks meant the City should promote two black captains, two black district chiefs and one black deputy chief. As to the deputy chief's position, Judge McRae reiterated the decree meant 20 percent promotions in each classification, including district chief and deputy chief. The judge pointed out that one person slightly more qualified was no basis for not meeting the goal specified in the decree.

Louis Britt, assistant City attorney, said the City was moving to meet the goal of 20 percent black promotions. He stated the problem was that black firefighters were impatient. He said, "I think what the plaintiffs want to do now is reach long-term goals immediately." Attorney Britt expressed the idea that black firefighters were only a short time away from promotions in all ranks. He continued, "If there are no eligible blacks (now), that should not be held against us." On the other hand, Attorney Fields felt the City had exhibited a foot-dragging attitude since the consent decree was approved in 1980. I feel that the City has exhibited a foot-dragging attitude toward blacks in the City of Memphis Fire Department since the very first black firefighters of April 1874.

The court insisted that the City of Memphis comply with the 1980 consent decree. Judge McRae ordered two black firefighters, rather than one, promoted to district chief. Of the thirty-three promotions, eight were black. These included District Chiefs Floyd E. Newsum and Carl W. Stotts and Captains James Arnold Jr. and Herbert Redden. Lieutenants included Danny Elliot Sr., Joseph H. Hodges and Eddie Newsom. I had all the qualifications, but it was the ruling of Federal Judge Robert McRae Jr. on June 19, 1982 that permitted me to become the first black deputy chief of the Memphis Fire Department.

Chapter 14

Unacceptable Compromises

As deputy fire chief, I was responsible for half of the city, all firefighting equipment, firefighters and battalions in that part of the city. Some of my responsibilities included commanding Division I. This division consists of five fire battalions. I conducted analysis of fire problems in the division and developed strategies for correcting problems. Other responsibilities included, but were not limited to, inspecting fire stations' apparatus and firefighters to make certain fire division standards were met, establishing emergency-scene safety, attending seminars for training and assuring that personnel were treated with dignity, fairness and without prejudice. I also conducted administrative investigations, hearings and disciplinary sanctions. Among other duties and responsibilities, I counseled firefighters in the department on personnel matters. I performed various other duties and responsibilities as required by the position and the fire department.

I'd follow up on fire calls anywhere in my division. Firefighters didn't know when or where I would appear on the fire scene. Therefore, for me, the fire scene offered an excellent opportunity for observing officers and firefighters in action. Observing firefighters in action, I felt better prepared to evaluate firefighters' performances more fairly. Firefighters handling fire hoses correctly, officers making the proper calls and everyone practicing fire safety were some areas of performance that I graded.

I learned early through reading, and confirmed through experience, that firefighting is much more than knowing where to point a water hose and putting water on fire. Firefighting is a discipline. Firefighting is a science that requires knowledge and skill.

I signed all evaluations on the men and their officers in my division. However, I found that some white officers overrated white firefighters and

City of Memphis fireboat. *Photograph courtesy of the Memphis Fire Department.*

underrated black firefighters. I paid particular attention to this as I signed the evaluations. If I felt that a firefighter was overrated, I'd question the supervisor and explain why I felt that person was overrated. Occasionally the evaluation was changed. Other times I refused to sign an inflated evaluation. Sometimes I had problems with firefighters regarding evaluation, especially when they felt that they deserved a promotion for reasons other than being productive.

I met challenges very early in the position of deputy fire chief. One thing was certain; we all knew that I was not their choice. I was aware that I was not welcomed in the position. Therefore I was guarded, territorial and especially observant. My initial challenges as deputy fire chief occurred during my first staff meetings with the other five deputy fire chiefs. The purpose of that meeting was for the deputy chiefs to assign newly promoted firefighters to various stations.

It's understandable that deputy fire chiefs and district fire chiefs want the best men in their battalion and their companies. Consequently, deputy fire chiefs share the lists of promotions with the district chief in each battalion. The district chiefs tell the deputy chief which men they want from the new list of promotions. The deputy chief tries to accommodate by honoring the districts chief's request. This deputy chiefs' meeting was to assign men to various companies. The intensity of that staff meeting remains clear in my mind.

We were given a list of vacancies and a promotional list of all the men a few days prior to the staff meeting. Because of what happened at the staff meeting, I suspected the deputies met prior to the scheduled meeting at fire headquarters. Of course I was not invited to that meeting. These people were friends and worked together for many years. They most likely had more meetings outside the department than they had in the department. Nonetheless, the decisions they'd made at their meeting were totally unacceptable to me.

All deputies had a certain number of openings on each shift. I learned that they had made all the assignments, including mine. They left me no choice, no request, no say-so, no input into who I wanted on my companies. I immediately voiced my objection to their actions.

I said forcefully, "No, no. This is not going to work. No, you're not going to tell me where to put my men." They listened as I spoke. No one attempted to interrupt me. I don't know what expression they saw on my face, but my demeanor must have told them they had done something wrong. I continued, "Now, you're not going to tell me where to place my men." The room was quiet; everyone just stared. I figured they probably thought since I was the new kid on the block, and a black one at that, I was a pushover. I was aware of the duties of the deputy fire chief prior to this promotion. Nonetheless, I was not surprised that plots would go on behind my back. I was aware that information would not be shared with me. I didn't need to be unnecessarily contrary or oppositional. On the other hand, I was confident enough to know when to be assertive. Even though I didn't know many firefighters on the list, I managed to get some of the men I wanted on my companies. Whatever else happened, I think I gained their respect after they learned that I would not be pushed around or accept whatever they dished out.

I'd worked enough, observed enough and read enough about the department to know something about its operation. I welcomed the challenges. But injustices abound when the "good old boy" network controls. Nevertheless, I continued looking for fair play, being recognized as an individual and a ranking member of the Memphis Fire Department. Some changes continued to be far away, hidden somewhere, perhaps in the future. I realized the changes that occurred since 1955. I also realized we still had much work to do. At this point in my career as deputy chief I was in a position where no black firefighter had been. Most importantly, in this position as deputy chief I knew I was in a position where I could continue helping with hiring and promoting other black firefighters. We still needed each other's support, particularly regarding hiring and promotions and unfair treatment.

When I was lieutenant, captain and district chief, Chief Dove was in charge of different review boards. This was a plus for me since he'd requested that I serve on the hiring and promotion review boards with him. This place on the boards gave me an opportunity to recommend black firefighters for hiring, promotions and transfers. Chief Dove seemed to respect my opinion because he knew that I would not recommend a firefighter, black or white, unless I felt the firefighter earned the recommendation. As deputy chief, I sat on the district chief board with Fire Chief Dove. Meanwhile Stotts, Wallace and Newsum recommended to me and I recommended to Chiefs Dove and Garner.

The Pioneers and I continued meetings to discuss ideas in helping with hiring and promoting black firefighters. Sometime we had names to put with where we wanted a black firefighter placed. I used every legitimate opportunity available to me to support black firefighters. As with any case, some black firefighters gave us reasons to hold our recommendations. Even so, I felt this is how the fire department operated through it history. Many times Stotts, Wallace and I spoke to administration together. Other times I approached the director alone. I remember approaching Director Smith in his office. I asked if he would consider promoting more black firefighters.

"No," he said. "I will promote 20 percent as stated in the consent decree. No more, no less."

It comes as no surprise that the fire department is traditionally rooted in politics and nepotism. I observed many white firefighters had sons, nephews, relatives and friends in the department. I kept this in mind as I read the history of the fire department, not only in Memphis but in many other cities as well. I am not condemning this practice in so many words. However, I must say that it stands to reason then that if we were not welcomed initially, it would be even more difficult getting promotions once hired in the department in the first place. Yet, if that's the way the department operates, then we must have someone in place to assist us in getting hired and promoted. We had no support in administration, we had no relatives in the department, we had no political ties with the department and we had no one in place to recognize and acknowledge our skills for the job. We had to fight the societal aspects, political aspects, nepotism and the "good old boy" network.

When black firefighters recommended sons, cousins, brothers or friends, the Pioneers and I pushed hard to get them hired and even harder to keep them once they were hired. We stressed that while we could help getting them, the real work had to come from them. For example, when a Pioneer's son came through I talked to him. He was impressive. I gave him information he needed in seeking a job with the department. I proved right by recommending him for employment. The young firefighter was a hard worker. He worked himself up in rank from private to captain. We worked to get them on; we didn't want to lose them once they were hired. I never recommended a man sight unseen. I made it a point to meet with the recommended person before I considered a recommendation to the board. I helped black firefighters because there was no "good old boy network" for them to go through.

There seemed to be nothing we could do about the political aspects of the job. You can't kill a bear every day; sometimes the bear wins. All the same, recommendations that I made were legal. I stayed within rules and guidelines of the department. I did nothing inappropriate. I tried to use the system to beat the system.

The district fire chief made recommendations to the deputy fire chief for transferring firefighters in his battalions. I transferred firefighters for various reasons. Since there were few black firefighters, I positioned them where they would have a chance for promotions. Often, I didn't tell the men why I was transferring them. Sometimes they objected to being placed in the eastern part of the city, or "way out." Little by little they began to understand why I'd placed them in certain areas. For instance, I placed black firefighters at different stations and on different equipment. A black firefighter was on the snorkel as a lieutenant under a captain. When I became deputy chief I brought him on my shift. I put him in charge because no black firefighters were in charge of a snorkel.

A practice that I'd observed was the process the deputy fire chiefs used when transferring firefighters. When it came to moving firefighters around or changing firefighters from one shift to another, the chiefs tried to place all black firefighters on my shift. Because most white chiefs wanted nothing to do with black firefighters, they did not want black firefighters on their shifts. Even though I accepted many black firefighters, a few remained on other shifts. Many black firefighters remaining with white chiefs seldom got promotions. The chiefs would ask me "When are you going to promote James?" I'd ask, "When are you going to recommend him?" In many cases I never received an answer or the recommendation from the chief. Chiefs seldom made recommendations for black firefighters while at the same time promoting white firefighters. I feel that most of the time they had no intentions of promoting black firefighters on their shifts. I had no authority to promote men from another chief's company. Nonetheless, I made certain that eligible blacks on my company were promoted every chance I got.

In addition to helping black firefighters I also helped white firefighters who were outside the "good old boy network." I remember a white firefighter who was a friendly, alert worker. He was his own man. He minded his business and was a rather quiet-mannered guy. This firefighter, a driver at one of the engine houses, was not in the "good old boy network." He was in the same predicament that I'd been in when I was a lieutenant. He was promoted to lieutenant and worked under a captain, as I'd done. Sullivan still did not have his command when I returned to that station as deputy fire chief. In my opinion, this lieutenant was a good firefighter who deserved his own company. I never regretted giving him his own company. He proved to be a good leader. I strongly felt that he was blackballed because he respected me and treated me as an officer when I was a lieutenant.

We were aware of traps or strategies sometimes used to get a black firefighter in trouble or fired, so we stressed following rules and regulations of the department. We stressed taking problems through proper channels and keeping records. As hard as we tried, we were not able to help every black firefighter. Some black firefighters ignored our advice. Some criticized our intentions, saying, for instance, they would not "beg" for promotions. I remember Wallace lecturing a Pioneer who displayed condescending attitudes toward black firefighters with higher ranks. "We need to respect each other," Wallace admonished. "How do you expect the white boys to act if you don't respect us?"

While Stotts, Wallace, the Pioneers and I worked in getting black firefighters hired and promoted, we also protected those we saw heading for trouble. We counseled them and tried to steer them in the right direction. At that time most of the black firefighters listened and appreciated our advice. Occasionally we'd come down hard on some black firefighter. We threatened to fire a few of them, but worked with them all because we cared.

Like other workplaces, the fire department was overloaded with gossip. Something was always swinging on the grapevine. Firefighters gossiped about everyone else, so I know they talked about me. When I made deputy chief, two deputies were assigned to each shift. Deputy Fire Chief E.B. Selph Division II and I was Division I. Deputy Chief Selph was our training officer in 1955. He was a lieutenant at that time. He was direct, he was steely and he took no mess, but he was a decent guy.

Deputy E.B. Selph and I frequently discussed different facets of the fire department. We also discussed different problems and solutions. For example, a district chief was a captain when I was a lieutenant at a fire station. We were promoted to district chief at the same time. He was still district chief when I was promoted to deputy chief. I became his division chief. He resented my promotion. That I became his superior and outranked him did not sit too well with him. I felt that I would have problems with this guy because of past and present situations. He was one of those who became bent out of shape because Kurt invited me to the Christmas party some years ago. This was also the same house that previously refused black firefighters' participation in the food syndicate some years ago.

This chief in command of a battalion, I'd been told, was dealing unfairly with black firefighters at the fire station. I'd talked to him concerning the way he treated black firefighters. He listened and said, "Ok, ok." However, as time passed, black firefighters in his battalion continued complaining that the district chief persistently treated them unfairly. The chief was doing things like ignoring blacks, intimidating them and displaying hostility toward them. The white firefighters simply followed the leader. During our discussion of the situation, Deputy Chief Selph said, "If I were you, I'd split up the whole house."

I said, "I need help in reassigning the men." The deputy chief gave me firefighters from his division. The four other deputy chiefs also helped. This was a double house with engine and truck companies. I transferred firefighters to other companies. The captain and lieutenant remained in the house with a new crew of firefighters. I knew separating the engine house was the best plan. However, the district chief didn't think so. Later, he stormed into my office bellowing ferociously. I felt and saw waves of his rage before he spoke.

With a scorching look, the chief shouted resentfully, "Why'd you break up my company?"

"Hold on," I snapped smartly. "You don't come in here raising hell with me. You sit down over there." He sat down and calmly tried to get it together.

He asked again, with less hostility, "Why'd you break up my house?"

"We had this conversation before and things got no better. You continued intimidating blacks on your company, you continue ignoring them, casting negative insults about black firefighters and talking negatively about them to your favorites in the engine house. You're not taking control of that house. You're not dealing with black firefighters in your battalion fairly. Your men, black and white, are not respecting each other. This is happening because of your attitude. In my opinion you are demonstrating poor leadership." I pointed out again what I expected of him, gave him a lecture about good leadership and attitudes of officers affecting the men. Needless to say the chief was furious when he left my office. Then again, part of my job was taking command. Taking command sometimes meant angering some firefighters.

A few months later, the chief returned to my office. A bit calmer, he requested that some of his men be returned to the engine house. "You have to have your head on right," I said to him. I lectured him again about his badmouthing black firefighters, playing favorites and demonstrating lack of control at the engine house. I returned some of his men. However, I didn't return the known troublemakers. It was no surprise that I would have to deal with a few "hardcores." I worked with them and worked with them. If I felt they were hopeless, I'd transfer them out. In such cases I always had support of one or two white deputies chiefs. When I'd transfer uncooperative white firefighters, I always explained where and why I was transferring them. Word was on the grapevine that "Deputy Chief Crawford won't hesitate to take action."

Once we got in position, things started to change, particularly in the department's hiring practices. Perhaps change would inevitably come in the department, but we were tired of waiting. By 1982 approximately 11 percent of the firefighters were black. We made some progress in the firefighting division as well as in other areas of the department. For instance, Sharlene Warren was the first female promoted to watch commander in the alarm office. Later, Virgie Watson won her case in court and was working in that same office as a dispatcher. She was later promoted to senior fire-alarm operator in the Fire Communication Bureau. She would later become watch commander in that department. Sharlene Warren worked in the alarm office. Norvell Wallace was president of the Credit Union Board of Directors. He was also chief in the Fire Prevention Bureau. Christine Campbell was manager of benefits in City personnel. Barbara Arnold, who worked in City personnel prior to 1955, and I were members of various credit-union committees. I was later chairman of one of the committees. Carl Stotts and Floyd Newsum were district fire chiefs. Affirmative action and the court were involved in almost all the advancement we made in the department. Since our problems were ongoing, we continued filing cases in court regarding discrimination and unfair treatment by administration and the department.

Generally speaking, the deputy fire chief position was a good position to be in if you planned to work longer. This position is not directly in the line of political action as with the mayor-appointed position of deputy director and the director of the department. At this point I had no intentions of seeking another position. I was ambitious to make things better. Then again, I didn't realize the unforeseen events that would force me to change my plans.

Chapter 15

The Future is Now

In 1985 Seattle, Washington, nine men and one woman accused of conspiracy to start a racist revolution were convicted on racketeering charges. Members of a white-supremacist gang called the Order were said to have committed crimes in a coast-to-coast spree to finance their plan to create a homeland free of Jews, blacks and white traitors. In Memphis, Sheriff Barksdale, frustrated by conflicting orders to end overcrowding at the Shelby County Jail, chained twelve convicted felons to a fence at the West Tennessee Reception Center. The famous Peabody Ducks appeared on the "Tonight Show." On the music scene, Tina Turner won a Grammy for the best record, "What's Love Got to Do With It?"

By 1985 many changes had occurred in my life. Delores and I now had five children. In addition to Joan and Antoinette, our family included Rosalind, Cassandra and Robert Jr. We had moved from a duplex we'd built on Philadelphia Street in Orange Mound. The street dead-ended at the railroad track. Fifteen seventy-four Wilson would be the first through street that we lived on. Our house, a three-bedroom red brick, was on a through street with similar homes, manicured lawns and friendly neighbors.

We lived at 1574 Wilson Street for ten years. After we outgrew that house we moved to a larger house at 1314 Fairmeade Avenue in Orchid Homes. This was one of many new developments of better homes for blacks in Memphis. Our new home was a two-story, four-bedroom brick and wood. A New Orleans type lacy-wrought-iron balcony crossed the upper front of the house. It was spacious. All the neighbors were new to the neighborhood. We became acquainted with many parents though our children. We had a neighborhood block club. Neighbors were cordial and whenever a neighbor or a member of one's family died, women in the neighborhood met at

Beale Street Historical District. The street is one of the main tourist attractions. Lined with nightclubs offering live music of jazz, blues, rock and roll, Beale Street is one of America's most famous streets. *Photograph courtesy of Robert Crawford.*

the home of the deceased with an assigned dish of food. At Christmas the women met and planned holiday decorations for the neighborhood. I remember our neighborhood won the first year we entered the city Christmas lighting contest. Many memories were made in this house.

Memories of our children remained long after they'd grown and moved away. Joan, the oldest, attended Chattanooga City College. She later attended Memphis State University. She married Earnest Washington. They have a son, Earnest Washington Jr. Joan works for City government in the comptroller's office.

Antoinette Janelle was married to Karl Turner. They have twin daughters, Karla and Stefanie. Antoinette received a master's degree in dance education from the Ohio State University. She is a Master Instruction Artist for the Kentucky Center for the Arts. She lives in Louisville, Kentucky, with her husband, James Willis.

Our daughter, Rosalind Elaine, opted to attend vocational school. She is a welder for Burlington Northern Railroad. Our last daughter, Cassandra, earned degrees from University of Tennessee, Knoxville, and Xavier School of Pharmacy. She married Gerald Taylor. They have three daughters, Rachael, Mallory and Peyton.

There is an interesting story behind Robert Jr.'s career. Robby attended Subiaco Academy in Searcy, Arkansas. He later attended and finished Hamilton High School in Memphis. He accompanied me on various jobs. He was quick to catch on. He liked taking things apart to see how they worked. Sometimes he'd reassemble an object on his own. Other times he'd seek my advice when working on a project. He was exceptionally adept at building models cars, airplanes and boats. Because he didn't seem too interested in higher education, I refrained from nagging him about his career. I felt he needed more time to think about what he wanted to do with his life. I'd planned to be supportive in whatever he wanted to do. However, Delores felt he'd had twelve years to make a decision. She decided to assist him in finding himself. I was surprised at what she'd done.

Robby had grown accustomed to watching late-night TV. He woke up one morning and found a sign facing him. Delores had written in bold letters: You have three options. (1) Go to school (2) Get a job (3) Join the army.

Two weeks later Robby invited Delores and me to his swearing-in ceremony. He found his niche in the U.S. Coast Guard. He enlisted and served six years in the coast guard. Our son died of pneumonia in December of 1995.

Delores earned a master's degree from the University of Tennessee School of Social Work. She had worked as a nurse and later as a social worker at the Department of Human Services. She later worked as a clinical social worker at the Hemophilia Clinic at the University of Tennessee. Working with staff and patients at this clinic was apparently satisfying, as she appeared happier with that job than any of the others. Working with people has always seemed easier for her.

My wife and I experienced many of the same problems on the job as many other blacks living and working, in Memphis in particular and America in general. Regarding the workplace, we shared new experiences every day. She has always been supportive in every way.

I continued working two jobs. Floyd Newsum and I cut grass days we were not at the fire department. We later operated a neighborhood grocery store out of Newsum Annex on Dunnavant Street. We also ventured into real estate. We purchased, refurbished and sold houses. I worked as a journeyman plumber at Hill Plumbing Company. I continued in business long after I'd left the fire department. I opened Pizza on Beale Street Pizza Parlor and Mrs. C's Ice Cream Parlor on Beale Street. Our children worked the business. The grandchildren worked and learned about the business during school vacations and summers. After earning a Tennessee Builders License, I became president of R&D Development Co., LLC. I knew I'd want to keep busy after retiring from the fire department. I took the advice I'd been given long ago. "If you want to insure having a job, then open your own business." Although race relations in the fire department were not much different than the fifties, I found some positives, especially economically. However, struggles with the department were ongoing.

I experienced few problems with subordinate staff. We treated each other respectfully. Firefighters invited us to lunch at various stations. I believe that seeing black and white working together served as positive models for most men at the stations. Deputy Chief Jack Jenkins and I worked especially well together. Jack, a portly man, was quiet and serious about his work. However, we like many people in Memphis, avoided discussing racial issues. Nonetheless, I feel that Jack was his own man.

In spite of the struggles and the discriminatory cases filed for injustices by the Pioneers, the fire department continued to drag its feet in hiring black firefighters. Fear of competition and the "good old-boy network" legacy will apparently remain part of the system. I have heard the fear of competition from some white firefighters that "Crawford is moving too fast."

That I was "moving too fast" was a problem for some firefighters. However, it appeared the real problem was the ingrained beliefs of die-hards that black firefighters were inferior, that we were never qualified to perform any job and therefore not eligible for employment or promotion with the fire department. From what I have observed, at times fear of competition seemed stronger than other traits, such as justification by religion.

A newspaper was running a competition to discover the highest principled, sober, well-behaved local citizen. Among the entries came one that reads: "I don't smoke, drink or gamble. I am hard working, quiet and obedient. I never go to movies or the theater, and I go to bed early every night and rise with the dawn. I attend chapel regularly every Sunday without fail. I've been like this for the past three years. But just wait until next spring, when they let me out of here."

Rumor that Director Smith planned to retire reached the grapevine long before it was officially announced. I remember a conversation with the director before he retired. While having coffee one morning, the director asked why I didn't apply for the position. I hadn't given the idea much thought until later. However, after the director's retirement became official, rumors started that I had plans to apply for the position. The rumor had it that I would probably get the promotion. The rumors had some white firefighters concerned, a few black firefighters annoyed and most Pioneers Black Firefighters hopeful. I offered no comments. I listened to rumors with apathy.

A park in the Beale Street area of Memphis, known as W.C. Handy Park, has a statue dedicated to W.C. Handy. *Photograph courtesy of Robert Crawford.*

The indifference about applying for the position was altered by one rumor in particular that angered and motivated me to act. The easiest way to anger a firefighter is to criticize his firefighting skills. I was aware that black firefighters had been accused, before we were hired in 1955, of being afraid of fire and therefore would not make good firefighters. I was aware of this psychological game used against me by a white firefighter long ago. I was fed up with this lie. Now I knew this was to deter my applying for the appointment. I flashed back to Director Smith's question regarding applying for the director's position. Without discussing with anyone and without hesitation, I applied for that position.

After I'd applied for the Director's position, events started following each other like days of the week. I was unaware when I applied for the Director's position the Pioneers had approached the mayor on my behalf. According to the Norvell Wallace, here is what happened:

> I was the first black deputy fire marshal promoted under the consent decree, was approached by captain Herbert Redden, who wanted to know why the Pioneers were not pushing somebody for deputy director's position since one of the functions of the Pioneers was to support minorities. After the discussion it was decided the Pioneers would support Crawford for the position. The Pioneers talked to the mayor regarding appointing Crawford as deputy director and about hiring minorities. At this time we had blacks in many positions in the department, but no black in higher positions. Robert Young, Jimmy Arnold, a few other Pioneers and I applied for that position for Crawford.
>
> We were surprised at what we heard when we talked to the mayor about the position. The mayor said that he'd learned that Crawford couldn't fight fire and that Crawford had lost two fires. Supposedly, Crawford lost a building on Thomas Street because he couldn't handle a multiple-alarm fire. Well, we said, he's been handling multiple-alarm fires. The mayor then asked about a fire in South Gate Shopping Center on Third Street that Crawford supposedly lost.
>
> We explained that those were the chief officer's fires. The officer on the scene was responsible for those fires and Crawford was not the chief officer on the scene. I know for a fact those were not Crawford's fires. I was the deputy fire marshal on duty those days. I was there.
>
> Our argument was based on facts, on Crawford's career with the City of Memphis Fire Department. How could he possibly come through the ranks and make it this far if he couldn't fight fire? He fought fires and more. Crawford went through the ranks from private, driver, lieutenant, captain, district chief, and is presently deputy chief. He was a fire instructor

who supervised firefighters. He investigated fires and could tell you how they started and where they started. He'd investigate a fire and tell what the company did right and what they did wrong. He has also attended various training seminars and workshops on firefighting. There's no way a Memphis firefighter, especially a black firefighter, could make ranks as Crawford did and not be able to fight fire. No way. We asked how anyone could claim that Crawford couldn't fires.

We've always been aware of the lies about us being afraid of fire still being spread by some white firefighters. Most district chiefs appreciated black firefighters' skills. I have seen district chiefs like Perry standing in the street waving us in at a fire. Some white firefighters did not like Chief Perry. They said he was too mean. We made runs with Chief Perry. He may have been cantankerous, but when we finished putting out a fire, he'd always tell us we did a good job. I have seen him bent over gagging, snot running everywhere. He'd see us coming and wave us on and say "Go in there and put that fire out for me."

White deputies knew this, which is why we couldn't understand a chief saying Crawford couldn't fight fires. The Pioneers and I concluded that the officer either wanted the job, didn't want to work with Crawford or both. We were angry when we learned about the lies told about Crawford.

If black firefighters trained by "experts"—and I especially, having gone through the ranks—could not fight fire and were unskilled firefighters, then the glory of Memphis Fire Department as one of the best in the nation was nothing more than a miserable joke. This mind game that blacks couldn't fight fire was no less the same kind of mind game spread during World War II meant especially for the women in Europe. "Stay away from black soldiers," goes the myth, "they have tails like monkeys that come out at night." As with the black soldiers of that time, lies about black firefighters rooted in fear of competition continued echoing from those who saw us as a threat even prior to 1955.

Director J. R. Smith retired from the department May 17, 1985. I did not get the director's position. Mayor Hackett appointed District Chief Billy G. Burross, Director of Fire Services. Replacing Deputy Chief B.G. Hall who recently retired, I was appointed Deputy Fire Director, the highest-ranking black officer in the history of the Memphis Fire Department.

Chapter 16

Disjointed Alliances

I have been active in various organizations all of my life. I'd opened two businesses at 341 Beale Street, Mrs. C's Ice Cream Parlor and Pizza on Beale. In addition to spending time with family, I continued activities with the church and other organizations as well. I was chairman of the Pastoral Council and a member of the Long Range Planning Committee at St. Augustine Catholic Church. In church-related organizations, I was a member of the Board of Directors of Associated Catholic Charities and treasurer of the St. Augustine-Father Bertrand Alumni Association. I served on the advisory committee of the City of Memphis Credit Union. I observed and learned much about organizational structure and operations through my years of actual participation. Thus, I knew what constituted good management, good leadership and good teamwork. The deputy director position, however, proved to be the most challenging.

My office at headquarters was equipped with an oversized desk, two chairs, a couch, end tables, lamps and a large plant. The chair behind the desk faced the office entrance. My back was against the murky Mississippi River. Large prints with Memphis scenes hung on each wall. I opted to keep the secretary, since she was knowledgeable about the system. A good secretary who knows the organization is an asset to an administrator. Since I was new, it made little sense to have a new secretary.

There was no training for the deputy director's position. I was given no instruction, and it was apparently assumed that I knew nothing about the job. Consequently, I was offered an officer "to help you make decisions." I responded, "I don't need help with making decisions. We will discuss at meetings," I said, "but the final decision is mine." The officer offered to me just happened to be a rank below me and with less experience on the job.

Man of the year, awarded by Men's Club of St. Augustine Catholic Church. *Photograph courtesy of Father John Geaney, CPS.*

I'd observed previous actions of directors and deputy directors. I was well acquainted with the policy manual long before I was appointed to the position. Basically, I knew very well what the job was about. Therefore, I wasted no time in doing what I had to do. I was never inclined to idle.

The deputy director's duties included ensuring that the bureaus and section assigned to command staff operated efficiently. I submitted annual comprehensive reports of activities of the firefighting and training bureaus. I was responsible, through commanding officers, for extinguishing fire and protecting lives and property at fires and other emergencies. I also had authority to take all appropriate and necessary action in the event of fire and other emergencies. I represented the director of fire service upon request or in his absence. I was responsible for recommending to the director of fire service for submission to the mayor and city council, the addition of fire station's personnel and equipment to provide adequate protection of life and property within the city. I'd observed and learned much, negative and positive, from previous administrators. Coming through the ranks certainly had rewards.

Like most Pioneers, I knew rejection. That some firefighters would reject me was guaranteed. Some individuals made their feelings clear about a black deputy director before I was appointed. In any case, it had been the department's history. Blacks were not wanted in any positions and certainly not commanding positions. Although no one wants to be rejected, it doesn't always have to be a bad thing. I felt the important thing was to stay focused and make rejection work in a positive way for me. The pressure I felt was to break through the "good old boy network," as I was at a place where I could be fair with all personnel, black and white. All blacks and some white firefighters were excluded from the network.

The administrators didn't want a black deputy director, no matter who he may have been. Even though they said little, their behavior made their feelings clear. They were the type of leaders who would have done anything to block the appointment. It was a long time before they made eye contact during conversations. They'd shift their eyes in all direction, avoiding eye contact. Their action was to make me feel worthless, useless. They used their behavior as weapons. I would not let their perception of me become my perception of me. I didn't buy it coming through the ranks and I wasn't buying it now. These were individuals. I was fortunate to have the experience of dealing with unbiased whites growing up and on the job as a firefighter. They spoke for themselves.

In the past, the top team leaders were the three principal managers of department affairs. I'd received training in leadership perception, organization and planning. In this administration the second-ranking

Memphis Queen Line Riverboats. Authentic paddlewheeler cruises on America's mighty Mississippi year-round. Cruises include sightseeing day and night trips. The cruise has floating gift shops and restaurants. *Photograph courtesy of Robert Crawford.*

Grizzlies presents award during Black History Week. *Photograph courtesy of Memphis Grizzlies.*

team member was excluded, therefore, by department standards, our administration had no team because by design all players, though in place, were not included as viable team members.

I didn't appreciate the rejection. I could hear my father's voice saying something like, "Don't let it burden you. Challenge it." I did just that. I let the rejection make me strong so that I became so fully engaged in my duties; their behavior ceased getting my full attention. I let the rejection help me grow in other ways. I let it give me more reason to be there.

The administrators and I had meetings with the entire staff. Each of us had our own agenda. However, the only times I met with department leaders were in cases involving the consent decree or union contracts. Administration seldom talked with union representatives or dealt with questions regarding the consent decree. I was particularly familiar with both. I'd had labor relation and management training dealing with unions. I also helped draft changes in the local 1784 Union agreement. Therefore, I was very familiar with unions and union contracts. The fire department did business with the carpenters, firefighters, maintenance workers, painters and other unions. I was also summoned when there were problems regarding the consent decree. Small wonder that the administrators showed little interest in affirmative action. Of course I knew the consent decree through my work with the Pioneers.

I found myself interpreting or answering questions about the consent decree. It seemed some firefighters wanted to forget it. I would never let them forget the decree. In my opinion, administrators would have better served the department by showing interest in the consent decree and the unions, since both had serious issues in the past and both would be with the department for years to come. My complaint was not about having to deal with either the consent decree or the unions. Fire department personnel should have been familiar with the rules, regulations and procedures regarding unions and the consent decree and their impact on the ability of the department's mission. I refused to be limited in my functions as deputy director. Although I would never refuse to perform duties, I objected to my exclusion from other department business. I felt the leaders needed to have regular meetings to discuss strategies and to function as a team.

Administration was so tied up with denying my presence that they didn't know what I was doing. It was like looking back at the present. The game was to keep me in the dark regarding matters of the department. I felt ours was totally different from previous administrations I'd observed. In previous administrations, top leaders held meetings, discussed issues and made plans for the department and then met with subordinates. In my opinion, our administration was totally disconnected.

Any department is as good as its leaders. Consequently, what happens at the top filters to the bottom. Rumor was that morale was low in the fire department. Some firefighters complained there was a lack of leadership at the top. Some black firefighters made noise that "Crawford was not doing enough to get blacks promoted." Nonetheless, I continued to honor invitations to lunch at various fire stations. Even so, I was clearly aware of the rumors from black and white firefighters.

I discussed the situation with one of the administrators on more than one occasion. It was no secret that I'd applied for the director position. As we talked, something said confirmed my suspicion of being a threat to an administrator. My interest, I explained, was focusing on duties and responsibilities of the job to which I'd been appointed. When a member of the team looked bad, everybody looked bad. I didn't want to look bad. I am too much of a perfectionist to be satisfied with shoddy performances from others and especially from myself. The administrator gave me some crazy answer that had nothing to do with what I was talking about. We ended the meeting having accomplished nothing.

I felt that I was fighting a losing battle talking to the administration about attitudes and behavior. The team of which I was supposed to be a part planned no meetings. Therefore, we had no common agenda and no idea what each would cover with subordinate staff.

I held no great expectations of being accepted by some black or any white firefighters. I certainly didn't discount some hardnosed administrators. We are all products of our time. This was Memphis, 1985, with remnants of Memphis, 1955. Some still refused to accept blacks in commanding position. I remember vividly a meeting with the entire staff. During a discussion, everyone who offered an opinion was acknowledged except me. The attempt, evidently, was to deny my presence. As soon as I found a place to insert my opinion, I was interrupted by one of the other leaders and ignored. Since I was intentionally ignored, I became rude, forcefully getting my point across. This was not the first time this happened. Again, I couldn't let this deliberate disregard of my opinion and my position pass.

I confronted the administrator's discounting my presence during meetings. I don't recall the exact content of the conversation. However, from that time on, the administrator acknowledged my presence and my position. The administrator listened more to what I had to say when we were one on one than he did when we were in a group. This attitude didn't surprise me.

Chapter 17

And Still We Stand

I knew the fire department policy manual. When I was District Chief I'd helped write policy and procedures for the Memphis Fire Department's Rules and Regulations Manual. The manual was not always referred to or followed, in my opinion, as it should have been. It seemed that some people made their own rules when it came to certain firefighters. I saw many things not done according to policy and procedure. Fairness regarding hiring and promotions were among my concerns.

I was particularly concerned about the treatment of some applicants by a few department heads. I permitted heads of the department to interview applicants for hiring and promotions. Mrs. Barbara Arnold, who was mature and responsible, worked in personnel prior to the consent decree. Following guidelines of the consent decree, I'd send Mrs. Arnold the position I wanted to fill. On one occasion, I needed a black person with plumbing experience included in the applicants for interview. I needed to fill the fire-hydrant repairman position. She found a black applicant with plumbing experience among the applicants. Heads of the department interviewed all applicants. The black applicant came in second. A white applicant came in first. I was sure from the discussion a white applicant would be selected. The white applicant was not the most qualified and he had no plumbing knowledge. The black applicant had plumbing knowledge that the job required.

I knew the network continued in full swing to keep blacks from employment and promotions. In addition to lacking required experience, the white applicant had no plumbing experience. He was a relative of one of the white firefighters. Another disqualifying factor the interviewing department head failed to find was that the white applicant had been fired from a previous job with the City. Therefore, when the application came to me, I rejected him because he had also been suspended several times for various infractions. I hired the black applicant because of his experience and his qualifications. I hired him because it was the right thing to do. This type of situation occurred time and time again. Blacks always came second, never first, no matter how qualified. This is the way the "good old boy network" continued working.

Another situation happened with fire department maintenance shop applicants. Many mechanics were eligible for a supervisor's position, but only a few whites and one black applied for the position. Mrs. Arnold sent names for my review. I checked the applicants' qualifications prior to submitting them to department head for interviews. Even though he had all the qualifications the job required, the black applicant came in second. He had experience as a mechanic prior to coming to the fire department. He'd worked for Memphis Transit Authority as a diesel mechanic. However, he had less experience with the City of Memphis Fire Department. Nonetheless, seniority still counted because he'd worked for another branch of the City. Therefore, he was qualified for the job. According to the consent decree, when qualifications were the same, the position should be given to a black person. In many cases some supervisors tried to ignore the consent decree and showed little interest in its interpretations. In this case, they also ignored the City of Memphis personnel manual.

The chief mechanic ignored the fact the white applicant he selected for the supervisor had an uncle working as a mechanic in that shop. According to City personnel and procedure manual, you cannot directly supervise a relative. The white applicant's uncle would have supervised him. I explained this to the department head after the interviews. It seems that black applicants always came in second place for promotions or employment, even though the consent decree mandated that 20 percent of promotions had to be black and 80 percent had to be white in each rank and division of the fire department. The "good old boy network" was alive and well and adversely affecting both blacks and unconnected whites. Situations similar to the ones cited above occurred time and time again, thus continuing the department's legacy.

I am certain that every black firefighter has a story to tell. Their stories will be much like mine. The only difference would probably be the way each handles situations. I was pleased with my job as deputy director, the second in command of the department and the biggest threat ever to administration. However, I was not pleased with their behavior. Likewise, they were not pleased with my being there. It was no surprise that I affected the entire department. Surprises for me did occur from time to time.

Only once did an administrator stun me. He confronted me about a decision I made that involved a black and a white firefighter, each of whom was on leave due to job-related injuries. I was updating the administrator about their situations, indicating that each had been informed of the requirement to get his doctor's permission to return to work. The white firefighter didn't present a physician's release permitting a return to work. Since the black firefighter had authorization from his physician, I allowed the black firefighter's return work. I explained that I was putting the black firefighter on light duty because of his on-the-job injury. After I'd explained this, the administrator's next question surprised and definitely attacked a nerve. He asked me why I didn't return the white guy to work with job-related injury. I explained again that the white guy did not have a release from the physician. In general, I saw myself as calm, composed and unruffled, but on this day I truly lost it and so did he.

The administrator asked, "Is your decision not to return the white firefighter to work a racial thing with you?" I was no longer calm. I was infuriated by his insinuation. I spoke with anger that seemed foreign to my usual way of handling situations. Everything that the administrator had done that suggested racism and things he'd neglected to do rushed to my mind. I lashed out with venom that I would never have admitted owning until that moment.

"How dare you ask me this," I said. At this point we were eye to eye. I continued, "You're more of a racist than I could ever be. It was you who put a white guy who had a non-job-related injury back to work on light duty. A firefighter who accidentally lost a finger, a non-duty injury," I shouted. "You should not have put that guy back to work rather than a firefighter who was injured in the line of duty. I feel this is racist!" I remember raising my voice at this point louder than before. The disagreement had escalated to the point that the secretary left her desk and closed the door.

The administrator said nothing. He sat there staring at me. Since I felt he'd always been cool toward me, it was difficult to determine the impact of my words. Apparently, our anger had intensified over time. Mine because of the years of experiencing rejections and delays in promotions because of my blackness and his because of his negative attitude toward blacks and resentment because of my appointment as deputy director. After that day, the administrator and I had little to say to each other. However, I felt the heated discussion helped both of us clear the air. We never broached the subject of racism again, and nothing changed.

Situations at headquarters didn't deter me from my responsibilities. I knew what I had to do. I didn't sit around waiting for administrators to confer with me. I had to make my own decisions in various situations. These were situations where I had no room for second-guessing decisions I made. Some situations called for immediate solutions where I had to call the shots

Memoriam to firefighters killed in the line of duty. *Photograph courtesy of Robert Crawford.*

Deputy Director Robert J. Crawford at Fire Headquarters. *Photograph courtesy of the Memphis Fire Department.*

and worry later about the decision being challenged. Some of my actions were challenged because the decision was unpopular with the general public. However, I didn't change decisions for fear the decision would be challenged. For example, during one of Memphis's coldest winters, we had a fire at a homeless shelter. I made a decision to close the shelter. I was called to the mayor's office to explain why I'd made the decision. I explained that I'd closed the house for fire code and safety violations. I had photos of the violation. The house remained closed.

On other occasions, my decisions were not challenged. For example, when St. Joseph Hospital's basement flooded, I made the decision to use fire department equipment to pump water from the basement so the hospital could remain open even though the department did not use fire equipment for such tasks. At certain times I had to take action and stick to my decisions.

Sometimes you have to take charge of taking charge. You have to be forceful with your commands. I remember one of the largest fires we had was at Mapco Refinery. The fire involved an industrial fuel tank. The department's foam unit played a significant role in fighting this blaze. I stood on a moat. The moat was surrounding another fuel tank. Directing the fire, I radioed down to the district chief telling where to direct the foam.

He radioed back saying "I told them where to direct the foam. They just won't do it."

I left the hill and went down to him. He repeated. "I told them where to direct the foam. They just won't do it."

I looked him dead in the eye. "Well," I shouted, "You get on their ass like I'm getting on yours." I think the district chief was angry with me, but he directed his anger at his men. He shouted at them the same way I'd shouted at him. They quickly obeyed his orders.

There are many people who were influential and inspirational to me in my unforgettable journey from altar boy to deputy director of the Memphis Fire Department. One memorable person is Sam Qualls Jr. Sam was light-skinned, tall, portly and a gentle man who loved children. He was director of Qualls Funeral Home on 479 Vance Avenue. He was always supportive of black firefighters. He'd sit on the bench in front of Station #8 and listen to our problems. He always provided much-needed encouragement and more. When we were housed in that isolated firehouse on Mississippi Boulevard, we were not allowed to watch television with the white captain and lieutenant in the captain's quarters. Sam Qualls donated a television to Station #8 for black firefighters. Because of his support, his encouragement and his generous nature, in 1975, we requested that a pumper be named for him. Later, the name on the pumper was changed to honor a firefighter killed in the line of duty. When a new firehouse replaced the old firehouse on Mississippi and Crump Boulevard, the Pioneers were instrumental in naming the new firehouse on Mississippi and Georgia for Sam W. Qualls Jr.

I had many mentors and many experiences, some good, some bad during my thirty-two years with the fire department. It is customary that high-ranking officers' names are placed on fire apparatus upon retirement. For the first time in the department's history, a black firefighter who came through the ranks has his name on a City of Memphis Fire Department's apparatus. I am proud to be honored that ladder truck #19 at Station #29 located at 2147 Elvis Presley Boulevard acknowledges my service to the City of Memphis Fire Department. I am proud to have been one of the first twelve black firefighters, the Pioneers, all of whom were role models for black boys and girls.

A good part of my firefighting career was spent looking forward. As I approached retirement, I looked around and saw the way the Pioneers had been working and how they had grown and made a difference. I will always be grateful to the Pioneer Black Firefighters for their confidence in me in the past, and particularly during the years when they were confronted with issues concerning me personally.

To me, the Pioneers was the greatest organization black firefighters ever had. My respect will always be theirs. I hope present black firefighters maintain a united effort to confront new challenges that arise. I hope black firefighters continue to have the necessary inspiration and enthusiasm to perform their duties effectively. I would hope that the black firefighters maintain their vigilance and monitoring of the consent decree. Above all, I hope that they listen and understand the history of how they got to the positions they now occupy.

The years were full and difficult. Busy they were and great were the burdens. I enjoyed the challenges and would not have had it any other way.

I agree with Oliver Wendell Holmes who wrote, "it is very grand to die in harness, but it is pleasant to have the tight and heavy collar lifted from the neck and shoulders."

However, once you start working it's hard to stop. The work of the black firefighters began long ago. The three black firefighters of 1874 pointed to the door, the twelve black firefighters of 1955 opened doors and now the present black firefighters' task is to be certain those doors never close.

Robert and Delores Crawford.

About the Authors

Robert J. Crawford Sr. was the first black driver, district chief, deputy chief and deputy director of the Memphis Fire Department. With each position he was the highest-ranking black firefighter within the City's fire department. He earned various certificates and awards for his firefighting skills. Robert retired as a thirty-three year veteran of the City's fire department.

Robert still lives in Memphis with his wife, Delores.

Delores A. Crawford, who collaborated with Robert, has published in *Writers on the River*, a local writer's club magazine. She has also published a poem, *Leaf in the Fall*, in *World of Poetry Anthology*.

Please visit us at
www.historypress.net